# Introduction to Polymer Physics

# Introduction to Polymer Physics

M. DOI

*Department of Applied Physics*
*Nagoya University, Japan*

Translated by
H. SEE

This publication was supported by a
generous donation from the Daido Life Foundation.

CLARENDON PRESS · OXFORD

1996

Oxford University Press, Walton Street, Oxford OX2 6DP
Oxford   New York
Athens   Auckland   Bangkok   Bombay
Calcutta   Cape Town   Dar es Salaam   Delhi
Florence   Hong Kong   Istanbul   Karachi
Kuala Lumpur   Madras   Madrid   Melbourne
Mexico City   Nairobi   Paris   Singapore
Taipei   Tokyo   Toronto
and associated companies in
Berlin   Ibadan

*Oxford* is a trade mark of Oxford University Press

Published in the United States
by Oxford University Press Inc., New York

Koubunshi Butsuri (Polymer Physics) by Masao Doi
Copyright © 1992, 1995 by Masao Doi
Originally published in Japanese by Iwanami Shoten,
Publishers, Tokyo 1992
Translation © Howard See, 1995

A catalogue record for this book is available from the British Library

Library of Congress Cataloguing in Publication Data
Doi, M. (Masao), 1948–
[Kōbunshi butsuri. English]
Introduction to polymer physics / M. Doi ; translated by H. See.
Includes bibliographical references and index.
1. Polymers.  I. Title.
QD381.D6413   1995      530.4′13–dc20       95–14969

ISBN 0 19 851772 6 (Hbk)
ISBN 0 19 851789 0 (Pbk)

Typeset by Keyword Typesetting Services Ltd
Printed in Great Britain by Biddles Ltd, Guildford & King's Lynn

# Preface

This book has been written as a textbook on polymer physics, an area that has undergone a great evolution over the past decades, owing to the new concepts introduced by S. F. Edwards and P. G. de Gennes. This evolution has been characterized by the advancement in our understanding of the entangled state of polymers, namely, concentrated solutions and melts. As a result, a framework has now been established to understand the static and dynamic properties of polymers in solutions, melts, and gels. The purpose of this book is to present this framework to graduate students in a concise and self-contained manner. The reader should have a knowledge of under-graduate-level statistical mechanics, but graduate-level topics, such as the theories of phase separation, fluctuations, and Brownian motion, are explained in the text.

This book was originally written in Japanese as the first part of the book *Polymer physics and phase transition dynamics* in the Iwanami modern phy-sics series. The second half was written by Professor Akira Onuki. Although we worked independently, our enthusiasm to write a text on the physics of complex mesoscale systems was shared. I thank him sincerely for giving me scientific stimulation and energy during the difficult period of writing. The manuscript of the original book was read by the students in my group, and their comments and questions were very useful in improving the text. Finally, I thank Professors S. F. Edwards and P. G. de Gennes for introdu-cing me to the new era of polymer physics.

*Nagoya University*                                                                     M. D.
July 1995

# Contents

# 1
# Properties of an isolated polymer molecule

A polymer is a large molecule made up of many small, simple chemical units, joined together by chemical reaction. For example, polyethylene ($CH_3 — (CH_2)_N — CH_3$) is a long chain-like molecule composed of ethylene molecules ($CH_2 = CH_2$), and DNA is an extremely long molecule made up of up to $10^7$ nucleotides. Giant molecules like these occur naturally in living organisms and are also synthetically produced to be used all around us, for example plastics, rubber, etc.

Most artificially produced polymers are a repetitive sequence of a particular atomic group, and take the form ($— A — A — A —$). The basic unit of this sequence is called the 'structural unit' or 'monomer unit', and the number of units in the sequence is called the degree of polymerization. A molecule is usually called a polymer if the degree of polymerization exceeds 100, and it is possible to have polymers containing over $10^5$ units. There are even naturally occuring polymers with the degree of polymerization exceeding $10^9$.

Materials composed of these very long molecules display properties that are completely different from materials composed of small molecules. Generally speaking, polymeric materials are very flexible, like rubber, and are easily formed into fibres, thin films, etc. In this book, we will see how the physical properties of polymers can be understood through statistical mechanics.

As with most other substances, in order to understand the properties of polymeric materials we must consider a large assembly of molecules. However, in the case of polymers, the molecules themselves are very large, and we need to use statistical mechanics to calculate the characteristics of even an isolated polymer. One way to investigate the properties of a single polymer is to place it in a very dilute solution, so that interactions between the polymers can be neglected. Experimentally, such dilute polymer solutions are used to determine the size and molecular weight of the molecule. In this chapter, we will theoretically investigate the properties of an isolated, single polymer chain in solution.

## 2   Properties of an isolated polymer molecule

## 1.1 The ideal chain

### *1.1.1 The random walk model*

A polymer molecule has many internal degrees of freedom, for example
the rotational freedom about each C–C bond in the polyethylene molecule,
and so it can take on many different configurations. Because of this high
degree of flexibility, we can picture a polymer chain as a very long piece of
string, as in Fig. 1.1. To study such a polymer chain, let us consider first
of all the simple model of Fig. 1.2, where we assume that the chain follows
a regular lattice. The portions of the polymer lying on the lattice points are
called 'segments', and the rods connecting the segments are called 'bonds'.
Let $b$ be the length of each bond and $z$ the coordination number of the
lattice.

Let us assume that there is no correlation between the directions that
different bonds take and that all directions have the same probability. In
this case, the configuration of the polymer will be the same as a random
walk on the lattice, and so the calculation we are about to perform can also
be applied to the statistical properties of random walks.

Consider the 'end-to-end vector' $\boldsymbol{R}$ joining one end of the polymer to the
other, the average length of which can be thought of as an indicator of the
extent of spreading out or size of the polymer. If the polymer is made up of
$N$ bonds, with $\boldsymbol{r}_n$ the vector of the $n$th bond, we have

$$\boldsymbol{R} = \sum_{n=1}^{N} \boldsymbol{r}_n. \tag{1.1}$$

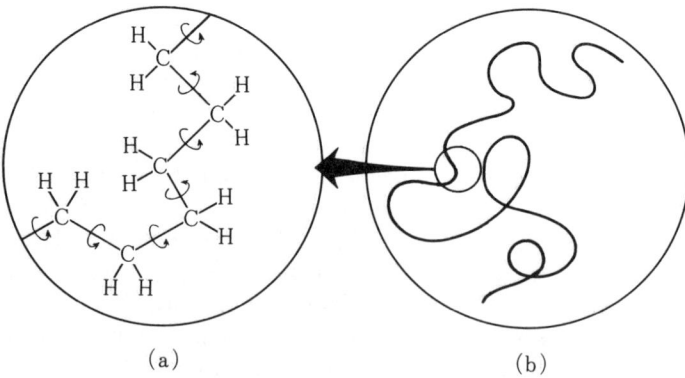

(a)                                                        (b)

**Fig. 1.1** (a) The atomic structure of the polyethylene molecule. (b) An overall
view of the molecule. There is rotational freedom about each C–C bond, so
the molecule as a whole resembles a long, flexible piece of string.

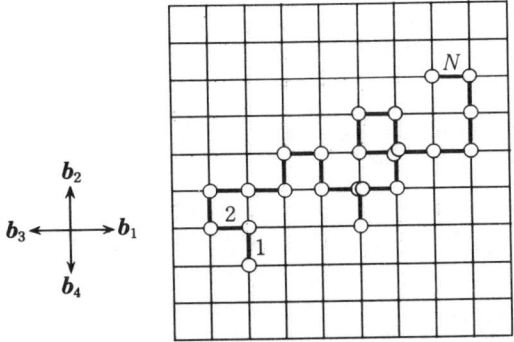

**Fig. 1.2** The random walk model of the polymer. The white circles are the segments and the thick lines are the bonds.

Clearly, the average value $\langle \boldsymbol{R} \rangle$ of $\boldsymbol{R}$ is zero, since the probability of the end-to-end vector being $\boldsymbol{R}$ is the same as it being $-\boldsymbol{R}$ so that the two contributions cancel out. Therefore we will calculate $\langle \boldsymbol{R}^2 \rangle$, the average of the square of $\boldsymbol{R}$, and express the size of the polymer by taking the square root of this quantity. From (1.1),

$$\langle \boldsymbol{R}^2 \rangle = \sum_{n=1}^{N} \sum_{m=1}^{N} \langle \boldsymbol{r}_n \cdot \boldsymbol{r}_m \rangle. \tag{1.2}$$

Since there is no correlation between the directions of different bond vectors, if $n \neq m$ then $\langle \boldsymbol{r}_n \cdot \boldsymbol{r}_m \rangle = \langle \boldsymbol{r}_n \rangle \cdot \langle \boldsymbol{r}_m \rangle = 0$. Therefore we find

$$\langle \boldsymbol{R}^2 \rangle = \sum_{n=1}^{N} \langle \boldsymbol{r}_n^2 \rangle = N b^2. \tag{1.3}$$

So we see that the size of the polymer is proportional to $N^{1/2}$.

It is also easy to calculate the probability distribution function of $\boldsymbol{R}$. Assume that we have a polymer of $N$ bonds, and that one end of the polymer is fixed at the origin. Let $P(\boldsymbol{R}, N)$ be the probability that the other end of the polymer is at a position $\boldsymbol{R}$. Writing $\boldsymbol{b}_i (i = 1, ..., z)$ for the possible bond vectors that the polymer can take, we see that if the polymer end has reached position $\boldsymbol{R}$ after $N$ steps, then the position vector at the $(N-1)$th step must be one of $\boldsymbol{R} - \boldsymbol{b}_i$, each of which has a probability of $1/z$ of occurring. Thus the probability of the polymer end being at $\boldsymbol{R}$ can be written as

$$P(\boldsymbol{R}, N) = \frac{1}{z} \sum_{i=1}^{z} P(\boldsymbol{R} - \boldsymbol{b}_i, N - 1) \tag{1.4}$$

If the polymer is very long, we have $N \gg 1$, $|\, R \,| \gg |\, b_i \,|$, and the right-hand side of the above equation can be expanded in terms of $N$ and $R$ as follows:

$$P(R - b_i, N - 1) = P(R, N) - \frac{\partial P}{\partial N} - \frac{\partial P}{\partial R_\alpha} b_{i\alpha} + \frac{1}{2} \frac{\partial^2 P}{\partial R_\alpha \partial R_\beta} b_{i\alpha} b_{i\beta}. \tag{1.5}$$

Here $b_{i\alpha}, R_\alpha$ are the components of $b_i, R$, and we have used the Einstein convention for summation over repeated indices. Substituting (1.5) into (1.4) and noting that

$$\frac{1}{z} \sum_{i=1}^{z} b_{i\alpha} = 0 \tag{1.6}$$

and

$$\frac{1}{z} \sum_{i=1}^{z} b_{i\alpha} b_{i\beta} = \frac{\delta_{\alpha\beta} b^2}{3} \tag{1.7}$$

yields

$$\frac{\partial P}{\partial N} = \frac{b^2}{6} \frac{\partial^2 P}{\partial R^2}. \tag{1.8}$$

Solving the differential equation (1.8) under the condition that $R$ is at the origin when $N = 0$ gives us

$$P(R, N) = \left( \frac{3}{2\pi N b^2} \right)^{3/2} \exp\left( -\frac{3R^2}{2N b^2} \right). \tag{1.9}$$

So we see that the probability distribution of $R$ is Gaussian. Actually, (1.3) and (1.9) are well-known results from the theory of random walks.

### 1.1.2 The effect of short-range interactions

In the model of the previous section, we assumed that the orientation of each bond is random and completely independent of the orientation of the previous bonds. This means that the polymer is able to fold back on to itself at certain locations, which is a physical impossibility since two portions of the polymer cannot occupy the same region in space. To remedy this, let us now consider a modified model of the polymer which disallows such doubling back. In the modified model, let us agree that the bond vector $r_{n+1}$ is not allowed to point back to the previous step, that is, it cannot take the direction $-r_n$, but must take one of the remaining $(z - 1)$ directions at random. Thus in this model the average value of $r_{n+1}$ will not be 0 for a given $r_n$. Writing this average as $\langle r_{n+1} \rangle_{r_n}$, we note that since

$$0 = \sum_{i=1}^{z} b_i = (z - 1)\langle r_{n+1} \rangle_{r_n} - r_n, \tag{1.10}$$

we have

$$\langle \boldsymbol{r}_{n+1} \rangle_{\boldsymbol{r}_n} = \frac{1}{z-1} \boldsymbol{r}_n. \tag{1.11}$$

Therefore, we find $\langle \boldsymbol{r}_{n+1} \cdot \boldsymbol{r}_n \rangle = b^2/(z-1)$. In the same way, we can calculate $\langle \boldsymbol{r}_{n+2} \cdot \boldsymbol{r}_n \rangle$. To do this, we first take the average of $\boldsymbol{r}_{n+2}$ for a fixed $\boldsymbol{r}_{n+1}$, giving us

$$\begin{aligned}
\langle \boldsymbol{r}_{n+2} \cdot \boldsymbol{r}_n \rangle &= \langle \langle \boldsymbol{r}_{n+2} \rangle_{\boldsymbol{r}_{n+1}} \cdot \boldsymbol{r}_n \rangle \\
&= \frac{1}{z-1} \langle \boldsymbol{r}_{n+1} \cdot \boldsymbol{r}_n \rangle \\
&= \frac{b^2}{(z-1)^2}.
\end{aligned}$$

Repetition of this process gives us the general result

$$\langle \boldsymbol{r}_n \cdot \boldsymbol{r}_m \rangle = \frac{b^2}{(z-1)^{|n-m|}}. \tag{1.12}$$

In this model, $\langle R^2 \rangle$ for a polymer chain is calculated as follows:

$$\langle R^2 \rangle = \sum_{n=1}^{N} \sum_{m=1}^{N} \langle \boldsymbol{r}_n \cdot \boldsymbol{r}_m \rangle = \sum_{n=1}^{N} \underline{\sum_{k=-n+1}^{N-n} \frac{b^2}{(z-1)^{|k|}}}. \tag{1.13}$$

If $N$ is very large, then for almost all $n$ the underlined summation can be replaced by one for $k$ from $-\infty$ to $\infty$, giving

$$\langle R^2 \rangle = \sum_{n=1}^{N} \sum_{k=-\infty}^{\infty} \frac{b^2}{(z-1)^{|k|}} = Nb^2 \frac{z}{z-2}. \tag{1.14}$$

Therefore, even with our modified model of the polymer, there is no change in the fundamental result that $\langle R^2 \rangle$ is proportional to $N$ for large $N$.

In general, if the interaction between the bonds extends only up to a finite distance along the chain, or in other words, if the total energy of the system can be written as

$$U_{\text{chain}} = \sum_n U(\boldsymbol{r}_n, \boldsymbol{r}_{n+1}, \ldots, \boldsymbol{r}_{n+n_c}) \tag{1.15}$$

then the quantity $\langle \boldsymbol{r}_n \cdot \boldsymbol{r}_m \rangle$ will decay exponentially for large $|n-m|$. (Actually, this is a general property of one-dimensional systems with finite interaction lengths). For such systems $\langle R^2 \rangle$ is always proportional to $N$ for large $N$, and the distribution of $R$ is Gaussian. In this sense, polymer models whose energy can be written in the form of (1.15) are equivalent to the random walk model, and such polymer chains are called 'ideal chains'. The average of the square of the end-to-end distance of an ideal chain can be written

$$\langle R^2 \rangle = Nb_{\text{eff}}^2, \tag{1.16}$$

where $b_{\text{eff}}$ is called the effective bond length. From now on, for simplicity we will write $b$ for $b_{\text{eff}}$.

Interactions which occur only between segments in close proximity along the chain, as in (1.15), are called 'short range interactions'. Note that here 'short range' refers to distances along the polymer chain, and not distances in space. (Actually, the interactions between polymers extend only over a range of nanometres, similar to small molecules.)

In actual polymer chains, two segments will interact if they happen to be spatially close, even if they belong to widely separated parts of the chain. These types of interactions, which depend only on actual spatial separations and not distances along the chain, are called 'long range interactions'. (Here 'long range' also refers to the distances along the polymer chain.) An example of a long-range interaction is the excluded volume interaction, which prevents any two segments from simultaneously occupying the same point on the lattice. As we shall soon see, the inclusion of long-range interactions causes a dramatic departure in the statistical properties of the chain from ideal chain behaviour. The main point here is that the ideal chain model takes into account the short-range effects, and ignores long-range interactions.

*1.1.3 Gaussian chains*

As we have seen above, for models that ignore long-range interactions the overall statistical properties of the chain do not depend on the details of the model if $N$ is large. Therefore, to obtain an overall description of the chain, it is convenient to use a model with as simple a mathematical formulation as possible, the lattice model of 1.1.1 being a good example.

Among non-lattice models of polymer chains, the Gaussian model is mathematically the simplest. This model assumes that the bond vector $\boldsymbol{r}$ itself possesses some flexibility and follows a Gaussian distribution:

$$p(\boldsymbol{r}) = \left(\frac{3}{2\pi b^2}\right)^{3/2} \exp\left(-\frac{3r^2}{2b^2}\right). \tag{1.17}$$

If we write the position vector of the $n$th segment in this Gaussian chain as $\boldsymbol{R}_n$, the distribution of the bond vector $\boldsymbol{r}_n = \boldsymbol{R}_n - \boldsymbol{R}_{n-1}$ is given by (1.17), and so the probability distribution of the set of position vectors $\{\boldsymbol{R}_n\} \equiv (\boldsymbol{R}_0, \boldsymbol{R}_1, \cdots \boldsymbol{R}_N)$ is proportional to

$$P(\{\boldsymbol{R}_n\}) = \left(\frac{3}{2\pi b^2}\right)^{3N/2} \exp\left(-\frac{3}{2b^2}\sum_{n=1}^{N}(\boldsymbol{R}_n - \boldsymbol{R}_{n-1})^2\right). \tag{1.18}$$

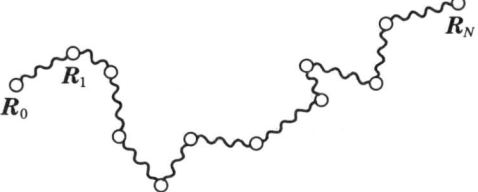

**Fig. 1.3** A Gaussian chain (the bead–spring model).

As shown in Fig. 1.3, we can think of the Gaussian chain as a linkage of segments consisting of harmonic springs of natural length 0. Letting $k$ be the spring constant, the energy of the chain can be written as

$$U = \frac{1}{2}k \sum_{n=1}^{N} (\boldsymbol{R}_n - \boldsymbol{R}_{n-1})^2. \tag{1.19}$$

The equilibrium state of this chain is described by a distribution function proportional to $\exp(-U/k_B T)$, and so if we choose the spring constant $k$ so that

$$k = \frac{3k_B T}{b^2}, \tag{1.20}$$

the chain's equilibrium distribution will be the same as (1.18). Because of this connection with springlike behaviour, the Gaussian chain is often called the bead–spring model.

## 1.2 Distribution of segments in the polymer chain

### 1.2.1 Pair correlation function

As we have seen above, the polymer chain occupies a roughly spherical region in space with a diameter of order $\langle R^2 \rangle^{1/2} = \sqrt{N}b$. Up to now we have only been concerned with the overall size of the polymer, but in this section we will investigate the spatial distribution of segments in the polymer chain. To do this, let us introduce the segment pair correlation function $g(\boldsymbol{r})$, defined as follows. Focusing attention on one segment (say the $n$th one), let $g_n(\boldsymbol{r})$ be the average density of segments at a position $\boldsymbol{r}$ from segment $n$. If we write $\boldsymbol{R}_n (n = 1, 2, ..., N)$ for the position vectors of the segments, then we can express $g_n(\boldsymbol{r})$ as follows:

$$g_n(\boldsymbol{r}) = \sum_{m=1}^{N} \langle \delta(\boldsymbol{r} - (\boldsymbol{R}_m - \boldsymbol{R}_n)) \rangle. \tag{1.21}$$

The pair correlation function $g(r)$ is the average of $g_n(r)$ over all $n$:

$$g(r) = \frac{1}{N} \sum_{n=1}^{N} g_n(r) = \frac{1}{N} \sum_{n=1}^{N} \sum_{m=1}^{N} \langle \delta(r - (\mathbf{R}_m - \mathbf{R}_n)) \rangle. \tag{1.22}$$

In addition, $g(\mathbf{q})$, the Fourier transform of $g(r)$, is

$$g(\mathbf{q}) = \int d\mathbf{r} e^{i\mathbf{q} \cdot \mathbf{r}} g(r) = \frac{1}{N} \sum_{n=1}^{N} \sum_{m=1}^{N} \langle \exp[i\mathbf{q} \cdot (\mathbf{R}_m - \mathbf{R}_n)] \rangle. \tag{1.23}$$

The quantity $g(\mathbf{q})$ can be measured experimentally by light scattering, small-angle X-ray scattering, etc.

### 1.2.2 Radius of gyration

From the behaviour of $g(\mathbf{q})$ at small $\mathbf{q}$, a length called the radius of gyration $R_g$ can be defined. Assuming $\mathbf{q}$ is small, (1.23) can be expanded with respect to $\mathbf{q}$ as follows:

$$g(\mathbf{q}) = \frac{1}{N} \sum_{n=1}^{N} \sum_{m=1}^{N} [1 - iq_\alpha \langle (\mathbf{R}_n - \mathbf{R}_m)_\alpha \rangle$$
$$- \frac{1}{2} q_\alpha q_\beta \langle (\mathbf{R}_n - \mathbf{R}_m)_\alpha (\mathbf{R}_n - \mathbf{R}_m)_\beta \rangle + \ldots] \tag{1.24}$$

Since the vector $\mathbf{R}_n - \mathbf{R}_m$ has an isotropic distribution, we can first take an average over the orientations. In general, if a vector $r$ has an isotropic distribution, the following relations hold for the average of its components:

$$\langle r_\alpha \rangle = 0,$$
$$\langle r_\alpha r_\beta \rangle = \frac{\langle r^2 \rangle}{3} \delta_{\alpha\beta}. \tag{1.25}$$

Using these, (1.24) becomes

$$g(\mathbf{q}) = \frac{1}{N} \sum_{n=1}^{N} \sum_{m=1}^{N} \left[ 1 - \frac{1}{6} q^2 \langle (\mathbf{R}_n - \mathbf{R}_m)^2 \rangle + \cdots \right] \tag{1.26}$$

If we write this as

$$g(\mathbf{q}) = g(0) \left( 1 - \frac{q^2}{3} R_g^2 + \cdots \right) \tag{1.27}$$

we have the following definition for $R_g$:

$$R_g^2 = \frac{1}{2N^2} \sum_{n=1}^{N} \sum_{m=1}^{N} \langle (\mathbf{R}_n - \mathbf{R}_m)^2 \rangle. \tag{1.28}$$

It happens that the radius of gyration $R_g$ is a more convenient way of expressing the size of a polymer than the average of the square of the end-to-end vector $\langle R^2 \rangle$. The radius of gyration can be directly measured in experiments, and can also be defined not only for linear chain polymers but also for polymers with branched structure, etc.

Notice that $R_g^2$ also equals the square of the average distance between the segments and the centre of mass of the polymer. The position of the centre of mass is defined by

$$R_G = \frac{1}{N} \sum_{n=1}^{N} R_n. \tag{1.29}$$

Using this, it is easy to show that $R_g$ can be rewritten as

$$R_g^2 = \frac{1}{N} \sum_{n=1}^{N} \left\langle (R_n - R_G)^2 \right\rangle. \tag{1.30}$$

### 1.2.3 Radius of gyration and pair correlation function of an ideal chain

Let us calculate the radius of gyration and segment pair correlation function for an ideal chain. When $|n - m|$ is large, $R_n - R_m$ of an ideal chain has a Gaussian distribution with variance $|n - m| b^2$, which means that

$$\langle (R_n - R_m)^2 \rangle = |n - m| b^2. \tag{1.31}$$

Therefore

$$R_g^2 = \frac{1}{2N^2} \sum_{n=1}^{N} \sum_{m=1}^{N} |n - m| b^2. \tag{1.32}$$

For large $N$ the summation can be replaced by an integration:

$$R_g^2 = \frac{b^2}{2N^2} \int_0^N dn \int_0^N dm \, |n - m| = \frac{1}{6} N b^2. \tag{1.33}$$

Therefore, the ratio of $R_g^2$ to $\langle R^2 \rangle$ is the constant $1/6$ for an ideal chain.

Further, $g(q)$ can be calculated in a similar way. Using the distribution function of $R_n - R_m$ we can write

$$\langle \exp[iq \cdot (R_n - R_m)] \rangle = \int dr \exp(iq \cdot r) \left( \frac{3}{2\pi |n - m| b^2} \right)^{3/2} \exp\left( -\frac{3r^2}{2 |n - m| b^2} \right)$$

$$= \exp\left( -\frac{q^2}{6} |n - m| b^2 \right). \tag{1.34}$$

Therefore

$$g(q) = \frac{1}{N} \int_0^N dn \int_0^N dm \exp\left( -\frac{q^2}{6} |n - m| b^2 \right) \tag{1.35}$$

$$= N f(q R_g).$$

Here

$$f(x) = \frac{2}{x^4}\left(\exp(-x^2) - 1 + x^2\right). \tag{1.36}$$

From (1.35) and (1.36), for $qR_g \gg 1$ we find

$$g(q) = \frac{2N}{q^2 R_g^2} \tag{1.37}$$

and for $q \to 0$ we have $g(q) = N$. The following Ornstein–Zernike type distribution function is a convenient interpolation between these two limits:

$$g(q) = \frac{N}{1 + q^2 \xi^2}. \tag{1.38}$$

Here $\xi^2 = R_g^2/2$. (1.35) and (1.38) agree to within 15% over all values of $qR_g$.


## 1.3 Non-ideal chains

### 1.3.1 The excluded volume effect

The ideal chain model only takes into account the short range interactions between segments which are located close to each other along the chain. Thus this model permits a chain to loop back onto itself so that segments which are widely separated along the chain will occupy the same region in space. Of course this is a physical impossibility since each segment possesses its own finite volume. In the lattice model of a polymer considered above, this effect can be accounted for by imposing the condition that two segments cannot occupy the same lattice site. In general, this type of condition is called the 'excluded volume effect'. If we model the polymer as a connected path on a lattice, the excluded volume effect will correspond to the condition that the path cannot pass through any sites that have been traversed previously. This is called a 'self avoiding walk', and the polymer thus represented is called an 'excluded volume chain'. An 'ideal chain' polymer corresponds to a random walk without the excluded volume effect.

The average size of an excluded volume chain is larger than that of an ideal chain, which can be easily seen as follows. For an ideal chain, there is a greater possibility of segments overlapping in a compressed polymer coil than in one of larger size. Thus when we add the restriction that no overlapping is permitted, we would expect the size distribution to be shifted to larger values, and so the excluded volume chain is larger than the ideal chain of the same length. Let us now estimate this effect by a simple model.

Consider the quantity $W(R)\mathrm{d}R$, which is the total number of excluded volume chains with the $N$th step lying at a distance between $R$ and $R + \mathrm{d}R$

from the origin. Since all possible paths have the same weight, $W(R)$ is proportional to the distribution function of $R$. In order to calculate $W(R)dR$, let us first ignore the excluded volume effect, and calculate the total number of ideal chains $W_0(R)dR$ with the $N$th step lying at a distance greater than $R$ but less than $R + dR$ from the origin. The overall number of ideal chains with $N$ steps is $z^N$, and the probability of the distance being between $R$ and $R + dR$ is $P(R, N)4\pi R^2 dR$, using the probability distribution function of (1.9). Therefore, we have

$$W_0(R) = z^N 4\pi R^2 \left(\frac{3}{2\pi Nb^2}\right)^{3/2} \exp\left(-\frac{3R^2}{2Nb^2}\right). \tag{1.39}$$

However, in the excluded volume chain, there are a number of ideal chain configurations which are disallowed due to the excluded volume effect. Let $p(R)$ be the probability that an ideal chain configuration, as counted in (1.39), is also allowable under the excluded volume condition. To estimate $p(R)$, assume that the polymer segments are evenly distributed in a region of volume $R^3$. If we write the volume of one lattice element as $v_c$, there will be $R^3/v_c$ lattice sites in the volume. We now calculate the probability that no overlaps occur when we place $N$ segments on these lattice sites, which will lead to an estimate for $p(R)$. The probability that one particular segment will not overlap with another is given by $(1 - v_c/R^3)$. Since there are $N(N-1)/2$ possible combinations of segment pairs, the probability that no overlap occurs in all of these combinations is given by

$$p(R) = (1 - v_c/R^3)^{N(N-1)/2} = \exp\left[\frac{1}{2}N(N-1)\ln\left(1 - \frac{v_c}{R^3}\right)\right]. \tag{1.40}$$

Since $R^3 \gg v_c$ we can estimate $\ln(1 - v_c/R^3) \simeq v_c/R^3$, and assuming $N \gg 1$ we obtain

$$p(R) = \exp\left(-\frac{N^2 v_c}{2R^3}\right). \tag{1.41}$$

Therefore

$$W(R) = W_0(R)p(R) \propto R^2 \exp\left(-\frac{3R^2}{2Nb^2} - \frac{N^2 v_c}{2R^3}\right). \tag{1.42}$$

The number of states $W(R)$ is proportional to the probability that the end-to-end distance of the excluded volume chain is $R$.

Both $W_0(R)$ and $W(R)$ have a maximum at certain values of $R$. Let us estimate the average size of the polymer chain in each model by calculating the positions of these maxima. The maximum of $W_0(R)$ occurs at $R_0^* = (2Nb^2/3)^{1/2}$. The maximum of $W(R)$ occurs at $R^*$, which satisfies the following equation obtained by differentiating the logarithm of (1.42):

$$-\frac{3R^{*2}}{2Nb^2} + \frac{3N^2 v_c}{4R^{*3}} + 1 = 0. \tag{1.43}$$

By combining these, we obtain

$$\left(\frac{R^*}{R_0^*}\right)^5 - \left(\frac{R^*}{R_0^*}\right)^3 = \frac{9\sqrt{6}}{16}\frac{v_c}{b^3}\sqrt{N}. \tag{1.44}$$

If $N \gg 1$, the second term on the left-hand side can be neglected, and then the solution to (1.44) is as follows.

$$R^* \simeq R_0^* \left(\frac{N^{1/2} v_c}{b^3}\right)^{1/5} \propto N^{3/5}. \tag{1.45}$$

We see that the characteristic size of excluded volume chains is proportional to $N^{3/5}$, and not $N^{1/2}$. The above is a very rough theory of the excluded volume effect. The statistical properties of excluded volume chains have been extensively investigated in numerical simulations, and for large $N$ it has been found that the size obeys the following formula:

$$R_g \simeq N^\nu b, \tag{1.46}$$

where the exponent $\nu$ is approximately 0.588, very close to the value $\frac{3}{5}$ calculated above.

### 1.3.2 Effect of the solvent

The models presented above do not explicitly take into account the influence of the solvent, whereas it is well known that the size of the polymer will greatly depend on the type of liquid in which it is placed. If there is a high affinity with the solvent so that the polymer is easily dissolved (this is called a 'good solvent'), the polymer configuration will be very spread out. On the other hand, in a solvent which does not dissolve the polymer (a 'poor solvent'), the polymer will be shrunken and compact. To explain this dependence of the polymer size on the type of solvent we must consider the interaction between the polymer and solvent molecules. In our lattice model, let us account for this effect in the following way.

For simplicity, assume that the solvent molecule is of the same size as a polymer segment and occupies one site on the lattice (see Fig. 1.4). Let us also assume that there are no voids, so that there are solvent molecules at all lattice sites not occupied by the segments. The interaction energies between neighbouring elements on the lattice are as follows:

$$\begin{array}{ll}
\textit{polymer segment} - \textit{polymer segment} & -\epsilon_{pp} \\
\textit{polymer segment} - \textit{solvent molecule} & -\epsilon_{ps} \\
\textit{solvent molecule} - \textit{solvent molecule} & -\epsilon_{ss}.
\end{array} \tag{1.47}$$

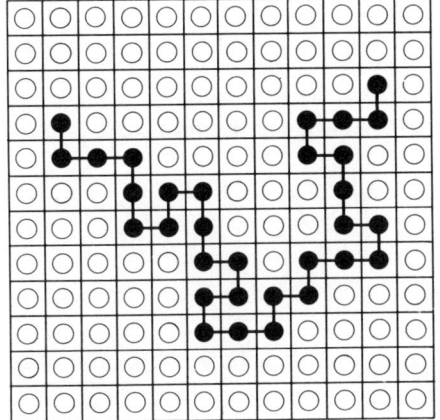

**Fig. 1.4** The lattice model of the excluded volume chain. The black circles are the segments of the polymer and the white circles are the solvent molecules.

These energies originate from the van der Waals attraction, and so $\epsilon_{pp}, \epsilon_{ps}, \epsilon_{ss}$ are positive.

For a given configuration $i$, let $N_{pp}^{(i)}$ be the number of neighbouring segment–segment pairs on the lattice, and let $N_{ps}^{(i)}$, $N_{ss}^{(i)}$ be the number of neighbouring segment–solvent pairs and solvent–solvent pairs, respectively. The overall system energy can then be written as follows.

$$E_i = -N_{pp}^{(i)}\epsilon_{pp} - N_{ps}^{(i)}\epsilon_{ps} - N_{ss}^{(i)}\epsilon_{ss}. \tag{1.48}$$

With these interactions present, the probability of an excluded volume chain having size $R$ is not proportional to $W(R)$, but can be written

$$P(R) \propto W(R) \exp\left[-\frac{\bar{E}(R)}{k_B T}\right] \tag{1.49}$$

Here, $\bar{E}(R)$ is the average energy of a polymer of size $R$.

As in the previous section, we assume that the polymer segments are uniformly distributed in a region of volume $R^3$, so that the probability that a lattice site in this region is occupied by a polymer segment is $\phi = Nv_c/R^3$. Therefore the average number of pairs $N_{pp}, N_{ps}, N_{ss}$ can be estimated as follows:

$$\overline{N_{pp}^{(i)}} \simeq \frac{1}{2}zN\phi,$$

$$\overline{N_{ps}^{(i)}} \simeq zN(1-\phi), \tag{1.50}$$

$$\overline{N_{ss}^{(i)}} \simeq N_{ss}^0 - [\frac{1}{2}zN\phi + zN(1-\phi)].$$

Here, $N_{ss}^0$ is the number of pairs of neighbouring solvent molecules when there is no polymer molecule in the system. Substituting (1.50) into (1.48) gives us the following approximation for $\bar{E}(R)$.

$$\bar{E}(R) \simeq -\frac{1}{2}zN\phi(\epsilon_{pp} + \epsilon_{ss} - 2\epsilon_{ps}) + \text{ terms independent of } \phi$$

$$= -\frac{zN^2 v_c}{R^3}\Delta\epsilon + \text{terms independent of } R. \tag{1.51}$$

Here $\Delta\epsilon$ is defined as follows:

$$\Delta\epsilon = \frac{1}{2}(\epsilon_{pp} + \epsilon_{ss}) - \epsilon_{ps}. \tag{1.52}$$

Substituting (1.51) into (1.49) shows us that the distribution function of $R$ has the same functional form as (1.42):

$$P(R) \propto R^2 \exp\left(-\frac{3R^2}{2Nb^2} - \frac{N^2 v_c}{2R^3}(1 - 2\chi)\right). \tag{1.53}$$

Here

$$\chi = \frac{z\Delta\epsilon}{k_B T} \tag{1.54}$$

is a non-dimensional quantity called the $\chi$ parameter. Comparing (1.53) and (1.42), we see that the effects of excluded volume and solvent interactions can be neatly expressed in terms of a single parameter

$$v = v_c(1 - 2\chi) = v_c\left(1 - \frac{2z}{k_B T}\Delta\epsilon\right), \tag{1.55}$$

where $v$ is called the excluded volume parameter. When there are interactions due to the solvent, in place of (1.44) the equation determining the size of the polymer $R^*$ is as follows:

$$\left(\frac{R^*}{R_0^*}\right)^5 - \left(\frac{R^*}{R_0^*}\right)^3 = \frac{9\sqrt{6}}{16}\frac{v}{b^3}\sqrt{N}. \tag{1.56}$$

### 1.3.3 The $\Theta$ temperature and coil–globule transition

The parameter $v$ includes not only excluded volume effects, but also the effects of solvent interactions as expressed by $\Delta\epsilon$. As shown in Fig 1.5, $\Delta\epsilon$ represents the decrease in energy when two polymer segments in the solvent come into contact. Therefore, if $\Delta\epsilon > 0$ the polymer segments will tend to come together, while if $\Delta\epsilon < 0$ they will tend to avoid each other. In most cases $\Delta\epsilon$ is positive, because the van der Waals forces, which are the main reasons for the attractive interactions, are proportional to the product of the electrical polarizabilities of the components. That is, if we

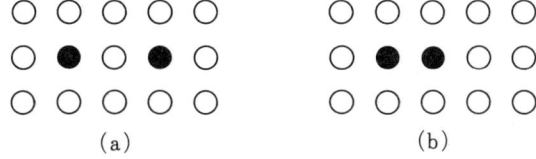

**Fig. 1.5** The effective interaction between polymer segments. If two polymer segments (represented by the black circles) are initially separated, as in (a), but are then brought together, as in (b), the energy of the system decreases by an amount $\epsilon_{pp} + \epsilon_{ss} - 2\epsilon_{ps} = 2\Delta\epsilon$.

write the respective polarizabilities as $\alpha_p, \alpha_s$, and the constant of proportionality as $k$, we have

$$\epsilon_{pp} = k\alpha_p^2, \quad \epsilon_{ps} = k\alpha_p\alpha_s, \quad \epsilon_{ss} = k\alpha_s^2. \tag{1.57}$$

Thus, substituting (1.57) into (1.52) shows us that

$$\Delta\epsilon = (k/2)(\alpha_p - \alpha_s)^2 > 0. \tag{1.58}$$

Therefore, the polymer segments tend to be in close proximity to each other under most circumstances.

In a good solvent, $\Delta\epsilon$ is small, and the excluded volume parameter $v$ is positive. On the other hand, in a poor solvent $\Delta\epsilon$ is large, and as the temperature increases $v$ will change sign from positive to negative at a certain temperature. The temperature at which the excluded volume parameter $v$ equals 0 is called the $\Theta$ temperature. From (1.55) the $\Theta$ temperature is given by

$$\Theta = \frac{2z\Delta\epsilon}{k_B}. \tag{1.59}$$

At the $\Theta$ temperature, the repulsive excluded volume effect balances the attractive forces between the segments, and the polymer behaves as an ideal chain.

Decreasing the temperature below the $\Theta$ temperature, the size of the polymer becomes much smaller than that of an ideal chain. According to (1.56) the quantity $R^*/R_0^*$ is determined not by $v$ but by $vN^{1/2}$, and so if $N$ is large only a small change in temperature will cause a big change in polymer size. For example, for polymers with a degree of polymerization of $10^6$, a variation of a few degrees in temperature will induce a dramatic change in the radius of gyration (see Fig 1.6). This change is called the coil–globule transition.

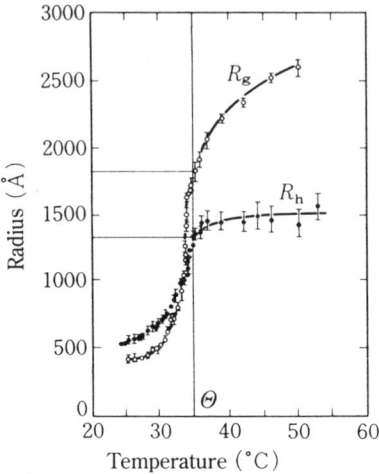

**Fig. 1.6** The coil–globule transition in a solution of polystyrene in cyclohexane. The radius of gyration $R_g$ and the hydrodynamic radius $R_h$ of the polymer show a dramatic change as temperature passes through the $\Theta$ temperature. The hydrodynamic radius $R_h$ is defined by $R_h = k_B T / 6\pi\eta D$, where $D$ is the diffusion constant of the polymer and $\eta$ is the viscosity of the solvent. (Sun, S.T., Nishio, I., Swislow, G., and Tanaka, T. (1980). *J. Chem. Phys.*, **73**, 5971, Fig.2.)

## 1.4 Scaling laws

The above theory of the excluded volume effect is based on the mean field approximation, because we have ignored the fact that the segments are connected together, and simply used the average segment concentration in our calculations. In actual fact, since the segments are linked there is a strong spatial correlation between them. To include the effect of these strong correlations, it is necessary to introduce the ideas of renormalization group theory, which was originally developed for the study of critical phenomena. It is beyond the level of the current text to discuss this theory, and the interested reader should consult the appropriate references listed at the end of the book.

Renormalization theory is very difficult, but among its conclusions there is a very simple and useful law. The idea of renormalization theory is to see how the macroscopic properties of the system change when the basic scales of the model are altered. Let us clarify this idea by considering the case of a Gaussian chain. For the Gaussian chain, the basic units are the segment length $b$ and the number of segments $N$. However, there is some arbitrariness in what we decide to call a 'segment'. As shown in Fig. 1.7, let us consider a group of $\lambda$ segments, and call this our new segment. The number

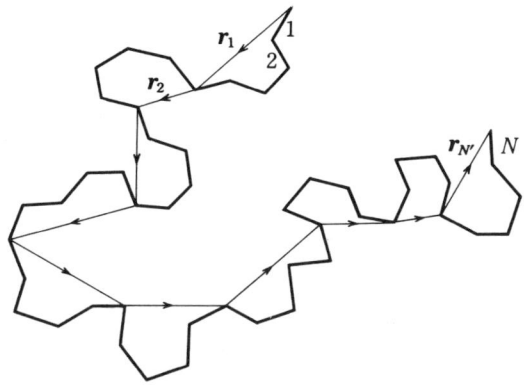

**Fig. 1.7** The original polymer consisting of $N$ segments (thick lines), and the new polymer made by linking only every fifth segment (arrows). The new polymer is made up from $N' = N/5$ segments, but the statistical properties of the polymer configuration do not change.

of new segments is $\lambda^{-1}N$ and the length of each new segment is $\lambda^{1/2}b$. In other words, $N$ and $b$ change as follows:

$$N \rightarrow \lambda^{-1}N \, , \, b \rightarrow \lambda^{1/2}b. \tag{1.60}$$

If $N$ is very large, there should be no change in the macroscopic properties of the polymer even after such a scale change. For example, the average of the square of the end-to-end distance $\langle \mathbf{R}^2 \rangle$ or the radius of gyration $R_{\mathrm{g}}$ are unaltered under the transformation of (1.60). Writing this mathematically, we say that these physical quantities satisfy the following equation:

$$f(\lambda^{-1}N, \lambda^{1/2}b) = f(N, b). \tag{1.61}$$

Using renormalization theory, it can be shown that there is a more general equation which also holds for chains with excluded volume:

$$f(\lambda^{-1}N, \lambda^{\nu}b) = f(N, b). \tag{1.62}$$

Here, the exponent $\nu$ is the same as that appearing in (1.46), and equals $\frac{1}{2}$ for an ideal chain, and approximately $\frac{3}{5}$ for an excluded volume chain.

In general for flexible polymers, when $N$ is large, the physical quantities which determine the overall properties of the chain satisfy the following relationship:

$$A(\lambda^{-1}N, \lambda^{\nu}b) = \lambda^{x}A(N, b). \tag{1.63}$$

Here $x$ is an exponent which depends on the physical quantity under consideration. This relation is called a scaling law.

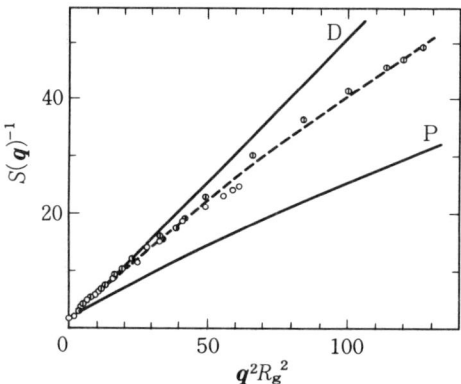

**Fig. 1.8** A plot of the inverse of scattered light intensity $S(\boldsymbol{q})$ against $\boldsymbol{q}^2 R_g^2$ for a solution of polystyrene in toluene. Data for polymers of different molecular weights are plotted: $\mathbf{\circ}$: $8.4 \times 10^6$; $\oplus$: $11 \times 10^6$; $\circ$: $21 \times 10^6$. (Noda, I., Imai, M., Kitano, T., and Nagasawa, M. (1985). *Macromolecules*, **16**, 425, Fig. 5.)

As an example of the application of a scaling law, let us consider the pair correlation function of the chain $g(\boldsymbol{q})$. From dimensional analysis, we can write

$$g(\boldsymbol{q}) = f_1(N, qb). \tag{1.64}$$

Under the transformation that $\lambda$ segments are grouped to form one segment, $g(\boldsymbol{q})$ will be reduced by $1/\lambda$, since $g(\boldsymbol{r})$ is proportional to the segment density, as in (1.22). Therefore

$$f_1(\lambda^{-1}N, \lambda^{\nu}qb) = \lambda^{-1}f_1(N, qb). \tag{1.65}$$

For this to hold true for arbitrary $\lambda$, the function $f_1(N, qb)$ must take the following form:

$$f_1(N, qb) = Nf_2(qN^{\nu}b), \tag{1.66}$$

where we have introduced a new function $f_2$. This can also be written as

$$g(\boldsymbol{q}) = Nf_2(qR_g). \tag{1.67}$$

Equation (1.67) tells us that if we were to measure $g(\boldsymbol{q})$ for polymers of different degrees of polymerization $N$, there would be superposition of the curves obtained by plotting $g(\boldsymbol{q})/N$ against $qR_g$ . This kind of scaling relation has been verified experimentally (see Fig. 1.8).

Scaling laws alone can only give us relations like (1.67), but if we further add some considerations of the physics involved, we can obtain even more useful conclusions. For $qR_g \gg 1$, the density correlation function should not depend on the length of the polymer chain, and so $g(q)$ should be independent of $N$. Since $R_g \propto N^\nu$, for (1.67) to be independent of $N$ we must have

$$g(q) = CN(qR_g)^{-1/\nu}, \tag{1.68}$$

where $C$ is a numerical constant. In other words, for $qR_g \gg 1$, we have

$$g(q) \propto q^{-1/\nu}. \tag{1.69}$$

In the case of a Gaussian chain, the relation (1.69) agrees with the calculation performed using (1.37). For excluded volume chains, (1.69) has been confirmed experimentally by light scattering.

# 2
# Concentrated solutions and melts

As the concentration in a polymer solution is increased, the molecules start to overlap and begin to entangle with each other as shown in Fig. 2.1. The critical concentration at which this commences is called the overlap concentration. Let us write $c^*$ for the number of segments per unit volume at this concentration, so that the number of polymers per unit volume is $c^*/N$. Since the volume of one polymer is of the order $R_g^3$, we must have

$$\frac{c^*}{N} R_g^3 \simeq 1. \tag{2.1}$$

As was explained in Chapter 1, $R_g$ is proportional to $N^\nu$, so we have

$$c^* \propto N^{1-3\nu} \simeq N^{-0.8} \quad \text{(for } \nu = 0.6\text{)}. \tag{2.2}$$

Notice that the overlap starts at very low concentration if $N$ is large. (For example, polystyrenes of molecular weight $10^6$ start to overlap at 0.5 % weight concentration.) Therefore, polymers with large molecular weight are almost always in the entangled state, and are strongly interacting with each other.

The limiting state of a polymer solution as concentration is increased is known as the polymer melt, which is a liquid state composed only of

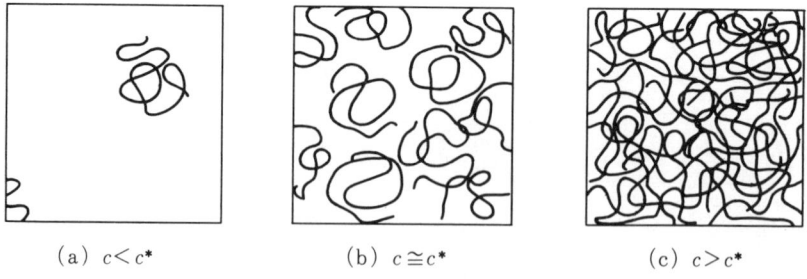

(a) $c < c^*$        (b) $c \cong c^*$        (c) $c > c^*$

**Fig. 2.1** (a) A dilute solution; (b) a solution at the overlap concentration $c^*$; (c) a concentrated solution.

polymers. This is an important state for industrial uses where polymeric materials are processed into various plastic products. Polymeric materials are often used with two or more components blended together, and the question of whether the polymers mix uniformly or separate into phases is an important one.

In this chapter, we will discuss the thermodynamic properties of overlapping polymers in solution, such as their phase separation, concentration fluctuations and osmotic pressure. Theories described here are based on a simple mean field approximation, but the fundamental characteristics of polymeric systems can be understood by seeing how the degree of polymerization $N$ enters the results.

## 2.1 Thermodynamic properties of polymer solutions

### 2.1.1 Flory–Huggins theory

Flory and Huggins proposed a simple theory to calculate the free energy of polymer solutions, which can be explained by the lattice model introduced in Chapter 1. In this model we assume that a polymer consists of $N$ polymer segments occupying $N$ connected lattice points, and that the rest of the lattice points are filled by solvent molecules. If we let $\Omega$ be the total number of lattice sites and $n_p$ the number of polymers in the system, then the number of solvent molecules $n_s$ is given by $n_s \equiv \Omega - n_p N$. To express the concentration of polymers, we shall consider the fraction of the lattice sites occupied by polymer segments,

$$\phi = \frac{n_p N}{\Omega}. \tag{2.3}$$

We call this the volume fraction.

The partition function of the system is given by

$$Z = \sum_i \exp(-E_i/k_B T), \tag{2.4}$$

where $i$ denotes a configuration (a way of arranging the $n_p$ polymers on a lattice) and $E_i$ is the energy of that configuration. To estimate $Z$, we replace $E_i$ by a constant value $\bar{E}$ independent of $i$. We now determine $\bar{E}$ by calculating the mean energy for the state in which polymer segments and solvent molecules are mixed randomly. Let $z$ be the lattice coordination number. Since, on average, each lattice point is surrounded by $z\phi$ polymer segments and $(1 - \phi)z$ solvent molecules, the number of neighbouring pairs of polymer segments is $N_{pp} = n_p N z\phi/2 = z\Omega\phi^2/2$. Similarly, the number of neighbouring pairs of solvent molecules is $N_{ss} = z\Omega(1 - \phi)^2/2$, and finally there are $N_{ps} = z\Omega\phi(1 - \phi)$ pairs of neighbouring polymer segment and solvent molecules. Denoting the interaction energy associated with these pairs as $\epsilon_{pp}, \epsilon_{ss}$ and $\epsilon_{ps}$, as in Section 1.3, we can write $\bar{E}$ as

$$\bar{E} \simeq -\Omega z \left[ \frac{1}{2} \epsilon_{pp} \phi^2 + \epsilon_{ps} \phi(1 - \phi) + \frac{1}{2} \epsilon_{ss}(1 - \phi)^2 \right]. \tag{2.5}$$

If $E_i$ is replaced by $\bar{E}$, $Z$ becomes

$$Z \simeq W \exp(-\bar{E}/k_B T), \tag{2.6}$$

where $W$ is the total number of allowed configurations for $n_p$ polymers.

To calculate $W$, let us consider the process of placing polymer chains on the lattice one after another. When we lay down the first polymer, the first segment can be put on $\Omega$ lattice sites, the second segment can be put on $z$ sites neighbouring to the first, and the third segment and those following can be put on $z - 1$ sites. Therefore the number of ways of placing the first polymer $w_1$ is given by

$$w_1 = \Omega z (z - 1)^{N-2}. \tag{2.7}$$

When $N$ is large, (2.7) can be approximated by

$$w_1 = \Omega(z - 1)^{N-1}. \tag{2.8}$$

Next, let us consider the number of ways $w_{j+1}$ of laying down the $(j + 1)$th polymer when $j$ polymers have already been placed. Since $Nj$ lattice sites are already occupied, $w_{j+1}$ can be estimated as

$$w_{j+1} \simeq (\Omega - Nj) \left[ (z - 1) \left( 1 - \frac{Nj}{\Omega} \right) \right]^{N-1} \simeq w_1 \left( 1 - \frac{Nj}{\Omega} \right)^N. \tag{2.9}$$

Therefore the total number of ways of placing $n_p$ polymers on the lattice is given by

$$W = \frac{1}{n_p!} \prod_{j=1}^{n_p} w_j. \tag{2.10}$$

The prefactor $n_p!$ is necessary because here we are assuming that the polymers are indistinguishable from each other.

Taking the logarithm of (2.10) and replacing the sum over $j$ by an integral, we have

$$\begin{aligned}
\ln W &= \sum_{j=1}^{n_p} \ln(w_j/j) \\
&= \int_0^{n_p} dj \left[ \ln \left( \Omega(z - 1)^{N-1} \left( 1 - \frac{Nj}{\Omega} \right)^N \right) - \ln j \right] \\
&= \Omega \left[ -\frac{\phi}{N} \ln \phi - (1 - \phi) \ln(1 - \phi) + \frac{\phi}{N}(1 + \ln N) + \phi \ln \left( \frac{z - 1}{e} \right) \right]. \tag{2.11}
\end{aligned}$$

If we substitute this into (2.6), we obtain the free energy $F = -k_B T \ln Z$ in the following form:

$$F = -k_B T \ln W + \bar{E}. \tag{2.12}$$

We will find it convenient to consider the free energy of mixing, $F_m$, which is defined as the free energy of the mixed state minus the sum of the free energies of the pure components. If the free energy of the solution as a function of the total number of lattice points $\Omega$ and polymer volume fraction $\phi$ is written as $F(\Omega, \phi)$, the free energy of the pure polymer can be written as $F(\Omega\phi, 1)$, and that of the pure solvent as $F(\Omega(1 - \phi), 0)$. Thus the free energy of mixing is given by

$$F_m(\Omega, \phi) = F(\Omega, \phi) - F(\Omega\phi, 1) - F(\Omega(1 - \phi), 0). \tag{2.13}$$

Substituting (2.5), (2.11), and (2.12) into (2.13), we obtain (after some calculation)

$$F_m = \Omega k_B T f_m(\phi), \tag{2.14}$$

where $f_m(\phi) k_B T$ is the mixing free energy per lattice site and is given by

$$f_m(\phi) = \frac{1}{N}\phi \ln \phi + (1 - \phi) \ln(1 - \phi) + \chi\phi(1 - \phi). \tag{2.15}$$

Here $\chi$ denotes the same quantity as that in Chapter 1:

$$\chi \equiv \frac{z}{2k_B T}\left[\epsilon_{pp} + \epsilon_{ss} - 2\epsilon_{ps}\right]. \tag{2.16}$$

### 2.1.2 Chemical potential and osmotic pressure

The free energy given by (2.14) and (2.15) is obtained for a fixed volume, and corresponds to the Helmholtz free energy. If we let $P$ be the pressure and $V = \Omega v_c$ the volume of the system, we can calculate the Gibbs free energy $G$ from $F$ as follows:

$$G(n_p, n_s, P, T) = F + PV = F + P(n_p N + n_s)v_c. \tag{2.17}$$

The chemical potential $\mu_s$ of the solvent is equal to the change of the Gibbs free energy when a solvent molecule is added to the system, keeping $n_p$, $T$ and $P$ constant. Thus

$$\begin{aligned}
\mu_s(\phi, P, T) &= G(n_p, n_s + 1, P, T) - G(n_p, n_s, P, T) \\
&= \left(\frac{\partial F}{\partial \Omega}\right)_{\phi,T}\left(\frac{\partial \Omega}{\partial n_s}\right)_{n_p} + \left(\frac{\partial F}{\partial \phi}\right)_{\Omega,T}\left(\frac{\partial \phi}{\partial n_s}\right)_{n_p} + Pv_c. \tag{2.18}
\end{aligned}$$

Using (2.14) and the equations $(\partial\Omega/\partial n_{\rm s})_{n_{\rm p}} = 1$, $(\partial\phi/\partial n_{\rm s})_{n_{\rm p}} = -\phi/\Omega$, we obtain

$$\mu_{\rm s}(\phi, P, T) = \mu_{\rm s}^0(T) + k_{\rm B}T\left(f_{\rm m} - \phi\frac{\partial f_{\rm m}}{\partial\phi}\right) + Pv_{\rm c}, \tag{2.19}$$

where $\mu_{\rm s}^0(T)$ is a function of temperature only.

Similarly, the chemical potential of the polymer segment is given by

$$\mu_{\rm p}(\phi, P, T) = \frac{1}{N}[G(n_{\rm p} + 1, n_{\rm s}, P, T) - G(n_{\rm p}, n_{\rm s}, P, T)]$$

$$= \mu_{\rm p}^0(T) + k_{\rm B}T\left(f_{\rm m} + (1 - \phi)\frac{\partial f_{\rm m}}{\partial\phi}\right) + Pv_{\rm c}. \tag{2.20}$$

Notice that the Gibbs–Duhem relation

$$(1 - \phi)\mu_{\rm s} + \phi\mu_{\rm p} = \frac{G}{\Omega}. \tag{2.21}$$

is satisfied for these equations.

We also see that the difference in chemical potential of a polymer segment and a solvent molecule is given by

$$\mu_{\rm p} - \mu_{\rm s} = \mu_{\rm p}^0 - \mu_{\rm s}^0 + k_{\rm B}T\frac{\partial f_{\rm m}}{\partial\phi}. \tag{2.22}$$

The osmotic pressure $\Pi$ is the extra pressure needed across a semi-permeable membrane to maintain the equilibrium of solvent molecules. Thus

$$\mu_{\rm s}(\phi, P + \Pi, T) = \mu_{\rm s}(0, P, T). \tag{2.23}$$

Substituting (2.19) into (2.23), and using $f_{\rm m}(0) = 0$, we have

$$\Pi = \frac{k_{\rm B}T}{v_{\rm c}}\left(\phi\frac{\partial f_{\rm m}}{\partial\phi} - f_{\rm m}\right). \tag{2.24}$$

For the free energy (2.15), this becomes

$$\Pi = \frac{k_{\rm B}T}{v_{\rm c}}\left[\frac{\phi}{N} - \ln(1 - \phi) - \phi - \chi\phi^2\right]. \tag{2.25}$$

For small values of $\phi$, eqn (2.25) can be expanded as a power series in $\phi$:

$$\Pi = \frac{k_{\rm B}T}{v_{\rm c}}\left[\frac{\phi}{N} + \left(\frac{1}{2} - \chi\right)\phi^2 + (1/3)\phi^3 + \dots\right]. \tag{2.26}$$

If $\phi \ll 1$, (2.26) becomes

$$\Pi = \frac{k_{\rm B}T}{Nv_{\rm c}}\phi = \frac{n_{\rm p}k_{\rm B}T}{V}. \tag{2.27}$$

This equation represents van't Hoff's law that the osmotic pressure of a solution is proportional to the number density of the solute molecules. Equation (2.26) indicates that in order for this law to be valid for polymer solutions, the concentration has to be very low. Indeed, the second term in (2.26) can be neglected only when

$$\phi \ll \frac{1}{(\frac{1}{2} - \chi)N}. \tag{2.28}$$

Since the right-hand side of this inequality is proportional to $1/N$, we see that for large values of $N$, the concentration for which van't Hoff's law is valid becomes very low.

Figure 2.2 shows a typical experimental result, with the osmotic pressure divided by the concentration plotted against the concentration. In this case, the curve is horizontal in the region where van't Hoff's law holds, and clearly this region becomes narrower as the molecular weight of the polymer increases.

If $N$ becomes very large, the first term in (2.26) can be neglected, so that

$$\Pi = \frac{k_B T}{v_c}(\frac{1}{2} - \chi)\phi^2. \tag{2.29}$$

Therefore the osmotic pressure becomes independent of the molecular weight for large $N$, which is also seen in Fig. 2.2.

According to (2.29), the osmotic pressure is proportional to the second power of concentration. Experimentally a slightly higher exponent is obtained. This deviation is now considered to be a failure of the mean field theory and will be explained later in Section 2.2.4. In any case, the characteristic feature of polymer solutions is that the osmotic pressure does

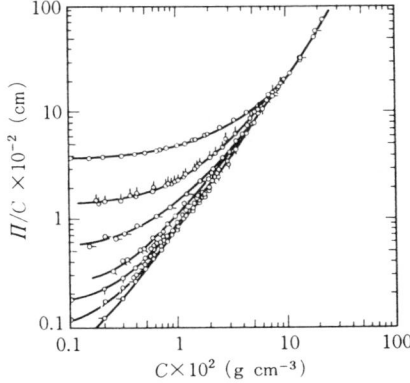

**Fig. 2.2** The concentration dependence of the osmotic pressure of poly($\alpha$-methyl styrene) molecules of various molecular weights dissolved in toluene. From the top, the molecular weights are $7 \times 10^4$, $20 \times 10^4$, $50.6 \times 10^4$, $7 \times 10^4$, $119 \times 10^4$, $182 \times 10^4$, $330 \times 10^4$, $747 \times 10^4$. (Noda, I., Kato, N., Kitano, T., and Nagasawa, M. (1981). *Macromolecules*, **16**, 668, Fig. 4.)

not obey van't Hoff's law even for concentrations as low as 1%. This observation, called the 'osmotic pressure anomaly,' attracted much attention in the early days of polymer science, but the reason is now well understood: it is because large polymers overlap with each other for even very dilute solutions.

According to (2.26), the coefficient of the $\phi^2$ term vanishes at the $\Theta$ temperature, which indicates a useful way to determine the $\Theta$ condition. Indeed, experimentally the $\Theta$ temperature is often defined as the temperature at which the second virial coefficient vanishes.

### 2.1.3 Phase separation

As was explained in Section 1.3, a poor solvent cannot accommodate many polymer molecules. Therefore when the polymer concentration is increased in poor solvents, the polymers will tend to aggregate, and beyond a certain concentration there will appear two phases, a phase of dilute solution and a phase of concentrated solution. This phenomenon is called 'phase separation'.

Whether the system remains homogeneous or separates into two phases can be predicted from the free energy of mixing $F_m(\phi)$. Suppose that the function $f_m(\phi)$ has the form shown in Fig. 2.3. Consider a system consisting of $\Omega\phi$ polymer segments and $\Omega(1 - \phi)$ solvent molecules. If the system remains in a homogeneous state, the free energy is given by $\Omega f_m(\phi)$, which corresponds to the point R in Fig. 2.3. Now suppose that the solution

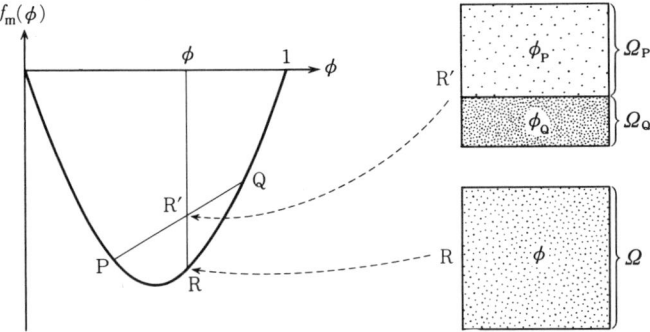

**Fig. 2.3** The shape of the mixing free energy curve for the case of no phase separation. State at $R$: concentration $\phi$, uniform phase of volume $\Omega$. State at $R'$: the previous state has now separated into two phases with concentrations $\phi_P$, $\phi_Q$ and volumes $\Omega_P$, $\Omega_Q$.

separates into two phases P and Q with volume fractions $\phi_P$ and $\phi_Q$, respectively. The volumes $\Omega_P$ and $\Omega_Q$ of each phase are determined from the following conditions of mass conservation and volume conservation:

$$\Omega_P \phi_P + \Omega_Q \phi_Q = \Omega \phi, \qquad (2.30)$$

$$\Omega_P + \Omega_Q = \Omega. \qquad (2.31)$$

Solving these gives us

$$\Omega_P = \frac{\phi_Q - \phi}{\phi_Q - \phi_P} \Omega \,, \; \Omega_Q = \frac{\phi - \phi_P}{\phi_Q - \phi_P} \Omega. \qquad (2.32)$$

Therefore the free energy of the phase-separated system is given by

$$F_m^{sep} = k_B T [\Omega_P f_m(\phi_P) + \Omega_Q f_m(\phi_Q)] \qquad (2.33)$$
$$= \Omega k_B T \left[ \frac{\phi_Q - \phi}{\phi_Q - \phi_P} f_m(\phi_P) + \frac{\phi - \phi_P}{\phi_Q - \phi_P} f_m(\phi_Q) \right].$$

This corresponds to the point R′ in Fig. 2.3. Therefore if the free energy curve is concave upwards for $0 < \phi < 1$ as in the case of Fig. 2.3, the free energy always increases when the system separates into two phases. In such a case there will be no spontaneous phase separation.

On the other hand, if $f_m(\phi)$ has two local minima as in Fig. 2.4, the system can lower its free energy by separating into two phases. For example, in the case shown in Fig. 2.4, the point R′ which represents the coexisting state of two phases of concentration $\phi_P$ and $\phi_Q$ has a lower free energy than that of the homogeneous state R. There are many possible ways of choosing P and Q, but the minimum of the free energy is attained when the line PQ coincides with the common tangent line in the figure. The corresponding concentrations $\phi_A$ and $\phi_B$ are determined from

$$\left[ \frac{\partial f_m}{\partial \phi} \right]_{\phi_A} = \left[ \frac{\partial f_m}{\partial \phi} \right]_{\phi_B} = \frac{f_m(\phi_B) - f_m(\phi_A)}{\phi_B - \phi_A}. \qquad (2.34)$$

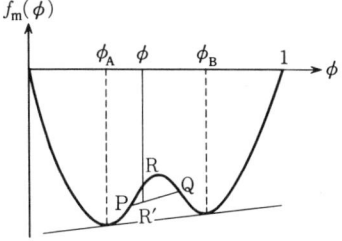

**Fig. 2.4** The shape of the mixing free energy curve for the case when phase separation occurs. For solutions with $\phi_A < \phi < \phi_B$, the free energy is lowest if the system separates into two phases of concentration $\phi_A$ and $\phi_B$.

It is easily shown that this condition is equivalent to the condition that the chemical potential of polymer (and solvent) in A phase equals that in B phase. This is easily checked using (2.19) and (2.20).

The shape of the function $f_m(\phi)$ varies with temperature. In the temperature range where $f_m(\phi)$ has two local minima, a solution with concentration $\phi$ ( $\phi_A < \phi < \phi_B$) will phase separate into two phases of concentration $\phi_A$ and $\phi_B$, as determined by (2.34). Such behaviour can be shown on a $\phi - T$ phase diagram as in Fig. 2.5. Here the gray region denotes the phase-separated state, and the remaining region is the homogeneous phase. The boundary between these two states is called the coexistence curve, and can be calculated from (2.34).

The extremum of the coexistence curve is called the critical point, and the corresponding temperature is called the critical temperature. The position of the critical point is obtained as follows.

In order to have double local minima, $f_m(\phi)$ must have an concave down region (ie. a region where $\partial^2 f_m/\partial\phi^2 < 0$) between $\phi_A$ and $\phi_B$. Therefore between $\phi_A$ and $\phi_B$ there must be two points C and D satisfying $\partial^2 f_m/\partial\phi^2 = 0$. The trace of such points in the $\phi - T$ plane is called the spinodal line. As the temperature approaches the critical temperature, the points C and D come closer, coinciding at the critical point. Since $\partial^2 f_m/\partial\phi^2 > 0$ above the critical temperature, the following equations have to be satisfied at the critical point:

$$\partial^2 f_m/\partial\phi^2 = 0 \ , \ \partial^3 f_m/\partial\phi^3 = 0. \tag{2.35}$$

For the free energy function of (2.15), the critical point is found to be

$$\phi_c = \frac{1}{1 + \sqrt{N}} \qquad \chi_c = \frac{1}{2}\left(1 + \frac{1}{\sqrt{N}}\right)^2. \tag{2.36}$$

Thus as $N$ increases, the critical concentration $\phi_c$ decreases and the critical temperature $T_c = z\Delta\epsilon/k_B\chi_c$ increases. Such a tendency is indeed observed experimentally, as in Fig. 2.6. However there is not good quantitative

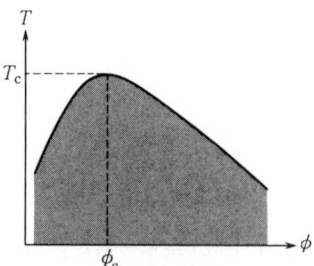

**Fig. 2.5** A phase diagram for a polymer solution.

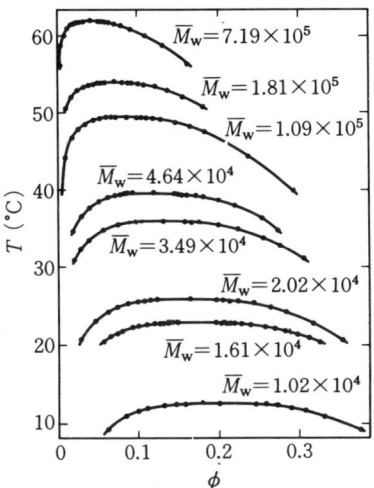

**Fig. 2.6** Coexistence curves for a solution of polystyrene in methylcyclohexane. Here $\phi$ is the volume fraction; the $\Theta$ temperature of this system is 70.3°C. (Dobashi, T., Nakata, M., and Kaneko, M. (1980). *J. Chem. Phys.*, **72**, 6692, Fig.1.)

agreement between the theoretical and experimental coexistence curves. The reason for the discrepancy is the large concentration fluctuations near the critical point, and to account for such effects, a theory beyond the mean field theory is required.

## 2.2 Concentration fluctuation in polymer solutions

### 2.2.1 Correlation function and response function

So far we have been discussing inhomogeneities in polymer solutions on a macroscopic scale. However if we look at the system on a molecular scale, we can see inhomogeneities even in a uniform phase. For example, in dilute solutions the segment density is high inside the polymer coil region of radius $R_g$, but it becomes zero outside this region, which is obviously a large inhomogeneity in the distribution of polymer segments. If the polymer concentration is increased, the polymer coils begin to overlap, and the inhomogeneities in segment density become smaller. In a polymer melt, which is the upper limit of polymer concentration, there are no inhomogeneities in the segment density.

Such inhomogeneities can be studied by light scattering, small angle X-ray scattering, and neutron scattering, which all yield much useful information on the structure and thermodynamic properties of polymer solutions. Here we present a simple mean field theory which describes the concentration fluctuations in a polymer solution.

To make the argument general, let us consider a mixture of two polymers A and B, having degrees of polymerization $N_A$ and $N_B$, respectively. The case of a polymer in solution can be handled by setting $N_B$ equal to 1.

As in the previous section, we will use the lattice model. We assume that all the lattice sites are occupied by either A or B segments. Let $\phi_A$, $\phi_B$ be the overall volume fractions of each type of segment, with $\phi_A + \phi_B = 1$. For a given configuration of polymers on the lattice, we introduce the variables $\phi_A(r)$ and $\phi_B(r)$ to describe the local concentration: we specify that $\phi_A(r)$ is equal to 1 if the lattice point located at $r$ is occupied by an A segment and is equal to zero otherwise, with a similar definition for $\phi_B(r)$. We will consider an incompressible system, so that all sites are occupied by polymer segments, and the relation $\phi_A(r) + \phi_B(r) = 1$ is satisfied for all $r$. Letting $\langle ... \rangle$ denote an equilibrium ensemble average, then

$$\langle \phi_A(r) \rangle = \phi_A. \tag{2.37}$$

The deviation of the segmental density for each lattice site is defined by $\delta\phi_A(r) = \phi_A(r) - \phi_A$. The fluctuation is characterized by the correlation functions of $\delta\phi_A(r)$ and $\delta\phi_B(r)$ as follows:

$$S_{AA}(r - r') = \langle \delta\phi_A(r)\delta\phi_A(r') \rangle \,, \; S_{AB}(r - r') = \langle \delta\phi_A(r)\delta\phi_B(r') \rangle. \tag{2.38}$$

Since $\delta\phi_A(r) = -\delta\phi_B(r)$, we have

$$S_{AA}(r) = S_{BB}(r) = -S_{AB}(r) = -S_{BA}(r). \tag{2.39}$$

Therefore the concentration fluctuations of an incompressible system are characterized by the single correlation function $S(r) \equiv S_{AA}(r)$.

In order to calculate $S_{ab}(r)$ (where $a, b$ represent A or B), we use the following relation known from linear response theory. Let us consider weak external potentials $u_A(r), u_B(r)$ which act respectively on the segments of A and B polymers. The change in the system's potential energy is

$$U_{ext} = \int dr[u_A(r)\phi_A(r) + u_B(r)\phi_B(r)] \tag{2.40}$$

$$= \sum_{a=A,B} \int dr[u_a(r)\phi_a(r)]. \tag{2.41}$$

(For the lattice model, the right-hand side should be written as a sum over the lattice sites $\sum_r$, but here we write it in the integral form for simplicity in later calculations.) Under the external potential, the average of $\phi_a(r)$ will deviate from $\phi_a$. If the external field is small, the deviation

$\overline{\delta\phi_a(r)} \equiv \langle\phi_a(r)\rangle_{\text{ext}} - \phi_a$ can be written as a linear function of the external potential:

$$\overline{\delta\phi_a(r)} = -\sum_b \int dr' \Gamma_{ab}(r - r') u_b(r'), \qquad (2.42)$$

where $\Gamma_{ab}(r)$ is called the response function. It is related to the correlation function $S_{ab}(r)$ as follows:

$$\Gamma_{ab}(r) = \beta S_{ab}(r), \qquad (2.43)$$

where $\beta = 1/k_B T$.

To prove (2.43), let us write the intrinsic energy of the system as $U_0$, so that the equilibrium average $\overline{\delta\phi_a}$ in the presence of the external field can be expressed as

$$\begin{aligned}
\overline{\delta\phi_a} &= \frac{T_r \delta\phi_a \exp[-\beta(U_0 + U_{\text{ext}})]}{T_r \exp[-\beta(U_0 + U_{\text{ext}})]} \\
&= \frac{T_r \delta\phi_a \exp[-\beta(U_0 + U_{\text{ext}})]}{T_r \exp[-\beta U_0]} \frac{T_r \exp[-\beta U_0]}{T_r \exp[-\beta(U_0 + U_{\text{ext}})]} \\
&= \frac{\langle \delta\phi_a \exp(-\beta U_{\text{ext}}) \rangle}{\langle \exp(-\beta U_{\text{ext}}) \rangle}, \qquad (2.44)
\end{aligned}$$

where $\langle ... \rangle = [Tr... \exp(-\beta U_0)]/[Tr \exp(-\beta U_0)]$ denotes the equilibrium average when the external field is not applied.

For weak external fields, $\exp(-\beta U_{\text{ext}})$ can be approximated as $1 - \beta U_{\text{ext}}$. From (2.41) and (2.44), it follows that

$$\overline{\delta\phi_a(r)} = \langle \delta\phi_a(r) \rangle (1 + \langle \beta U_{\text{ext}} \rangle) - \langle \delta\phi_a(r) \beta U_{\text{ext}} \rangle = -\beta \sum_b \int dr' \langle \delta\phi_a(r) \delta\phi_b(r') \rangle u_b(r').$$

$$(2.45)$$

Comparing this with (2.42), we get (2.43).

In the case of an incompressible system we can use (2.39) and obtain from (2.45) the following:

$$\overline{\delta\phi_A(r)} = -\beta \int dr' S(r - r')(u_A(r') - u_B(r')). \qquad (2.46)$$

As an example, consider the situation where the spatial variation of $u_A(r), u_B(r)$ is very gradual, so that these can be considered constant over the range of length scales characterising $S(r)$. In this case, (2.46) can be approximated as follows:

$$\overline{\delta\phi_A(r)} = -\beta(u_A(r) - u_B(r)) \int dr' S(r - r'). \qquad (2.47)$$

On the other hand, if $u_A(r), u_B(r)$ are almost constant, $\overline{\delta\phi_A(r)} \equiv \overline{\delta\phi(r)}$ is determined from the condition of thermodynamic equilibrium, which is that the chemical potentials of A and B are constants independent of position. This is the same as requiring that the quantity $\mu_A - \mu_B$ is constant. In the absence of an external field, $\mu_A - \mu_B$ can be written as $k_B T \partial f_m / \partial \phi +$ *constant*, from (2.22). In the presence of an external field, the chemical potentials of A and B change by $u_A(r)$ and $u_B(r)$ respectively, so the condition for thermodynamic equilibrium becomes

$$k_B T \left( \frac{\partial f_m}{\partial \phi} \right)_{\phi = \phi(r)} + [u_A(r) - u_B(r)] = C \quad \text{(Constant)}. \tag{2.48}$$

Therefore, if $\overline{\delta \phi}$ is small, we have

$$k_B T \frac{\partial^2 f_m}{\partial \phi^2} \overline{\delta \phi(r)} + [u_A(r) - u_B(r)] = 0. \tag{2.49}$$

Thus

$$\overline{\delta \phi(r)} = - \left( k_B T \frac{\partial^2 f_m}{\partial \phi^2} \right)^{-1} [u_A(r) - u_B(r)] \tag{2.50}$$

Comparing this with (2.47) gives us

$$\int d r' S(r') = \left( \frac{\partial^2 f_m}{\partial \phi^2} \right)^{-1}. \tag{2.51}$$

Since the Fourier transform of $S(r)$ is defined as

$$S(q) = \int d r e^{i q \cdot r} S(r), \tag{2.52}$$

eqn (2.51) can be rewritten as

$$\frac{1}{S(q = 0)} = \frac{\partial^2 f_m}{\partial \phi^2}. \tag{2.53}$$

Now $S(q)$ is directly related to the light scattering intensity, and so the left-hand side of this equation can be determined from scattering experiments. If $S(q = 0)$ is known as a function of concentration $\phi$, eqn (2.53) can be integrated to give $f_m(\phi)$. (Here the boundary conditions are that $f_m(\phi) = 0$ at $\phi = 0$ and $\phi = 1$.)

### 2.2.2 Random phase approximation

Now let us calculate $S(r)$ using the mean field approximation. We will make use of (2.43). First of all, let us consider the case where the polymers A and B are placed on the lattice at random, without excluded volume effects or interaction energies. In this case, $\langle \delta \phi_A(r) \delta \phi_B(r') \rangle$ equals zero, but $\langle \delta \phi_A(r) \delta \phi_A(r') \rangle$ and $\langle \delta \phi_B(r) \delta \phi_B(r') \rangle$ are not zero since the segments of the polymers are linked together. For the Gaussian chain introduced in Chapter

1, the correlation functions for the concentration fluctuations can be expressed in terms of the pair correlation function $g(r, N)$ as follows:

$$\langle \delta\phi_A(r)\delta\phi_A(r') \rangle = \phi_A[g(r - r'; N_A) - \phi_A] \equiv S_{AA}^0(r - r'). \tag{2.54}$$

$$\langle \delta\phi_B(r)\delta\phi_B(r') \rangle = \phi_B[g(r - r'; N_B) - \phi_B] \equiv S_{BB}^0(r - r'). \tag{2.55}$$

In general, if an external field $u_A(r), u_B(r)$ is applied to this system, the resulting change in the concentration $\delta\phi_A(r)$ is given by

$$\overline{\delta\phi_A(r)} = -\beta \int dr' S_{AA}^0(r - r')u_A(r'). \tag{2.56}$$

Now, in reality there are interactions between the chains, which we will take into account through the mean field approximation. If the concentrations of A and B segments at position $r$ are respectively $\overline{\phi_A(r)} = \phi_A + \overline{\delta\phi_A(r)}, \overline{\phi_B(r)} = \phi_B + \overline{\delta\phi_B(r)}$ , the molecular fields acting on the segments are given by

$$w_A(r) = -z\left[\epsilon_{AA}\overline{\phi_A(r)} + \epsilon_{AB}\overline{\phi_B(r)}\right]$$
$$w_B(r) = -z\left[\epsilon_{BA}\overline{\phi_A(r)} + \epsilon_{BB}\overline{\phi_B(r)}\right]. \tag{2.57}$$

Further, there is the conservation of volume condition $\phi_A(r) + \phi_B(r) = 1$, which can be represented in the following potential form:

$$U_{excl} \equiv \int dr V(r)[\phi_A(r) + \phi_B(r)]. \tag{2.58}$$

Here $V(r)$ is a potential determined from the volume conservation condition. The mean fields acting on segments A and B are, respectively, $w_A + V$ and $w_B + V$, and so $\overline{\delta\phi_A(r)}, \overline{\delta\phi_B(r)}$ are given by the following:

$$\overline{\delta\phi_A(r)} = -\beta \int dr' S_{AA}^0(r - r')[u_A(r') + w_A(r') + V(r')] \tag{2.59}$$

$$\overline{\delta\phi_B(r)} = -\beta \int dr' S_{BB}^0(r - r')[u_B(r') + w_B(r') + V(r')]. \tag{2.60}$$

On the other hand, the constraining relation $\phi_A(r) + \phi_B(r) = 1$ gives

$$\overline{\delta\phi_A(r)} + \overline{\delta\phi_B(r)} = 0. \tag{2.61}$$

Equations (2.57), (2.59), (2.60), and (2.61) form a set of simultaneous equations for the unknowns $\overline{\delta\phi_A(r)}, \overline{\delta\phi_B(r)}, V(r)$. If the solution is expressed in the form of (2.42), the correlation function can be determined from (2.43). This type of approximation is called the random phase approximation.[1]

---

[1] The name random phase approximation is due to the fact that this approximation was originally used for the correlation between electrons in solid state electron theory. A more apt name in the present case would be the linear mean field approximation.

To solve the above equations, we will use the Fourier transform defined as follows:

$$\overline{\phi_{Aq}} \equiv \int d\mathbf{r} e^{i\mathbf{q}\cdot\mathbf{r}} \overline{\delta\phi_A(\mathbf{r})} = \int d\mathbf{r} e^{i\mathbf{q}\cdot\mathbf{r}} \overline{\phi_A(\mathbf{r})}. \tag{2.62}$$

Equations (2.59), (2.60), (2.61) take the following form:

$$\overline{\phi_A} = -\beta S_{AA}^{(0)} \left[ u_A - z\left(\epsilon_{AA}\overline{\phi_A} + \epsilon_{AB}\overline{\phi_B}\right) + V \right] \tag{2.63}$$

$$\overline{\phi_B} = -\beta S_{BB}^{(0)} \left[ u_B - z\left(\epsilon_{BA}\overline{\phi_A} + \epsilon_{BB}\overline{\phi_B}\right) + V \right] \tag{2.64}$$

$$\overline{\phi_A} + \overline{\phi_B} = 0. \tag{2.65}$$

Here for simplicity we have dropped the subscript $q$. Solving these gives us

$$\overline{\phi_A} = -\beta \left[ \frac{1}{S_{AA}^{(0)}} + \frac{1}{S_{BB}^{(0)}} - 2\chi \right]^{-1} (u_A - u_B). \tag{2.66}$$

Here $\chi$ is defined in a way similar to that in (2.16):

$$\chi = \frac{z}{2k_B T} [\epsilon_{AA} + \epsilon_{BB} - 2\epsilon_{AB}]. \tag{2.67}$$

From (2.46) and (2.66) the Fourier transform of the concentration fluctuations is given as follows:

$$\frac{1}{S(q)} = \frac{1}{S_{AA}^{(0)}(q)} + \frac{1}{S_{BB}^{(0)}(q)} - 2\chi. \tag{2.68}$$

Alternatively, from (2.54), (2.55) we have

$$S(q) = \left[ \frac{1}{\phi_A g(q, N_A)} + \frac{1}{\phi_B g(q, N_B)} - 2\chi \right]^{-1}. \tag{2.69}$$

The above approximation has used the completely random state as a base, and has estimated the effect of interactions through a perturbation calculation. Therefore, this model is not applicable to systems with strong correlation effects, for example a solution near $c^*$ where there are large fluctuations in the concentration. However, the accuracy of this approximation improves as the concentration increases, and it holds quite well for polymer blends.

*2.2.3 Concentration fluctuations in concentrated polymer solutions*

To apply (2.69) to polymer solutions, we just set $N_A = N, N_B = 1$. When $N_B = 1$, $g(q, N_B) = 1$ and so (2.69) becomes

$$S(q) = \left[ \frac{1}{\phi g(q, N)} + \frac{1}{1 - \phi} - 2\chi \right]^{-1}. \tag{2.70}$$

From (2.70) and (2.53) we have

$$\frac{\partial^2 f_m}{\partial \phi^2} = \frac{1}{N\phi} + \frac{1}{1-\phi} - 2\chi. \tag{2.71}$$

This completely agrees with the result obtained from (2.15). Therefore, the current theory is at the same level of approximation as the Flory–Huggins theory developed in Section 2.1.1. Using (1.38) for $g(q, N)$, eqn (2.70) gives us the following:

$$S(q) = \frac{S(0)}{1 + q^2 \xi^2}. \tag{2.72}$$

Here

$$S(0) = \left[ \frac{1}{N\phi} + \frac{1}{1-\phi} - 2\chi \right]^{-1}$$

$$\xi^2 = \left[ 1 + \phi N \left( \frac{1}{1-\phi} - 2\chi \right) \right]^{-1} \frac{Nb^2}{12}. \tag{2.73}$$

The quantity $\xi$ is called the correlation length of the concentrated solution. If $N \gg 1$ and $\phi \ll 1$, then $S(0)$ and $\xi$ can be written as

$$S(0) = \phi \left( \frac{1}{N} + \frac{v}{v_c} \phi \right)^{-1}, \quad \xi = \sqrt{\frac{b^2 v_c}{12\phi v}} \tag{2.74}$$

where we have used the excluded volume parameter $v = v_c(1 - 2\chi)$. With an increase in concentration or excluded volume, the scattering intensity $S(0)/\phi$ per segment decreases and the correlation length $\xi$ becomes shorter. This is because as the concentration or excluded volume is increased the average repulsion between the segments increases, and so there is a stronger tendency for the segment concentration to become uniform. This variation of scattering intensity with concentration has been observed experimentally (Fig. 2.7).

### 2.2.4 Scaling theory

As discussed in Section 2.2.2, the random phase approximation does not give accurate results for systems with large concentration fluctuations. For large molecular weight polymers, even if there is overlap between polymer chains, the overall concentration remains low and the concentration fluctuations $\langle \delta \phi^2 \rangle / \phi^2$ are very large. Such a solution is called semidilute. The mean field approximation cannot be used for semidilute solutions, and so calculations are based on renormalization group theory. In this section, we will use scaling concepts to see how the results of the previous calculations change

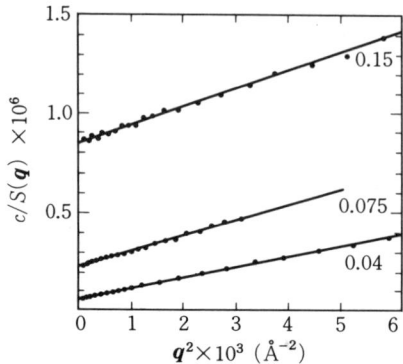

**Fig. 2.7** A plot of the inverse of the scattering intensity $S(q)$ against $q^2$ for a solution of polystyrene in carbon disulphide. The concentrations of the polystyrene are also indicated ($g\,cm^{-3}$). The corresponding correlation lengths are (from the top): $10\,\text{Å}$, $18\,\text{Å}$, $29\,\text{Å}$ (Daoud, M., Cotton, J.P., Farnoux, B., Jannink, G., Sarma, G., Benoit, H., Douplessix, R., Picot, C., and de Gennes, P.G., (1975). *Macromolecules*, **8**, 804, Fig. 10.)

for semidilute solutions. For simplicity, we will confine our discussion to polymers in good solvents.

*Osmotic pressure*

Using dimensional analysis, the osmotic pressure $\Pi$ can be written as a function of segment concentration $c$ and number of monomers in the polymer $N$ as follows:

$$\Pi = ck_B Tf(cb^3, N). \tag{2.75}$$

If we group $\lambda$ segments together and let this unit represent a new segment, $N, b, c$ will change as follows:

$$N \to \lambda^{-1}N \, , \, b \to \lambda^\nu b \, , \, c \to \lambda^{-1}c. \tag{2.76}$$

The pressure $\Pi$ is unchanged under this transformation, and so we have

$$\Pi = \frac{c}{N}k_B Tf\left(\frac{c}{N}(N^\nu b)^3\right). \tag{2.77}$$

At the concentration $c^*$ where the chains begin to overlap

$$c^* \simeq \frac{N}{(N^\nu b)^3}, \tag{2.78}$$

and the osmotic pressure becomes

$$\Pi = \frac{c}{N}k_B Tf\left(\frac{c}{c^*}\right). \tag{2.79}$$

This equation holds for both dilute and semidilute solutions. Assuming that the osmotic pressure is independent of $N$ for $c > c^*$, we must have

$$\Pi = \frac{c}{N} k_{\mathrm{B}} T \left( \frac{c}{c^*} \right)^{1/(3\nu-1)} \propto c^{9/4}. \tag{2.80}$$

This result is different to the mean field approximation result $\Pi \propto c^2$ of eqn (2.29). Experimental results are plotted in Fig. 2.2, and we see that the asymptotic slope of the curves is steeper than that predicted from the mean field approximation, but is close to the result of (2.80).

*Correlation length*

Using the same approach, it can be shown that the correlation length $\xi$ can be written as

$$\xi = R_{\mathrm{g}}^0 f \left( \frac{c}{c^*} \right). \tag{2.81}$$

Here $R_{\mathrm{g}}^0 \simeq N^\nu b$ is the radius of gyration of the polymer in dilute solution. If we use the fact that $\xi$ is independent of $N$ for $c > c^*$, we have

$$\xi \propto R_{\mathrm{g}}^0 \left( \frac{c}{c^*} \right)^{-\nu/(3\nu-1)} \propto c^{-3/4}. \tag{2.82}$$

The concentration dependence of this result is different to that of the mean field approximation (2.74) where $\xi \propto c^{-1/2}$. Experiments with semidilute solutions give results close to (2.82).

*The size of the polymer*

As we discussed in Chapter 1, in dilute solution a chain with excluded volume is more spread out than an ideal chain, since the overlapping of segments is minimized that way. However, in a concentrated solution where the polymers are jumbled together, if an individual chain spreads out there will not necessarily be a decrease in the overlapping of segments, and so the excluded volume effect does not act to spread out the chains. If the polymer concentration of a dilute solution is increased, the individual polymers decrease in size and at the concentration where there is sufficient interpenetration between the polymers, the chains will have the ideal chain configuration. This is called 'screening' of the excluded volume interactions. Scaling arguments can be used to estimate the dependence of the polymer size on the concentration.

As before, the radius of gyration of the polymer can be written as

$$R_{\mathrm{g}} = R_{\mathrm{g}}^0 f \left( \frac{c}{c^*} \right). \tag{2.83}$$

If $c \gg c^*$ the chain is ideal, and so $R_g$ is proportional to $N^{1/2}$. Therefore, we must have

$$R_g = R_g^0 \left(\frac{c}{c^*}\right)^{-(2\nu-1)/2(3\nu-1)} \propto c^{-1/8}. \tag{2.84}$$

Experimentally, the degree of spreading out of individual polymers in solution can be measured by the neutron scattering of specific polymers labelled with deuterium. The results of these experiments support (2.84).

## 2.3 Polymer blends

### 2.3.1 Phase diagrams of polymer–polymer mixtures

It is very difficult to make a uniform mixture of two different types of polymers. The reason for this can be explained using the Flory–Huggins theory. If we mix two different polymers A and B, with respective number of monomers $N_A, N_B$ and volume fractions $\phi_A, \phi_B$, the free energy of mixing can be written as follows:

$$f_m = \frac{1}{N_A}\phi_A \ln \phi_A + \frac{1}{N_B}\phi_B \ln \phi_B + \chi\phi_A\phi_B. \tag{2.85}$$

(This equation comes from integrating $\partial^2 f_m/\partial\phi_A^2 = 1/(N_A\phi_A) + 1/(N_B\phi_B) - 2\chi$ which is obtained from (2.69) and (2.53)). The last term in (2.85) represents the interaction energy, and since usually $\chi > 0$ this term acts to separate molecules of different type into phases. On the other hand, the first two terms represent the entropy of mixing, and these act to mix molecules of different types. For systems consisting of small molecules ($N_A = N_B = 1$) or polymers in solution ($N_A = N, N_B = 1$), the entropy of mixing terms are of order 1 and so overwhelm the interaction term, enabling the molecules to be mixed. However, in polymer–polymer mixes, the entropy of mixing is proportional to the inverse of the number of monomers and so becomes very small. Thus, unless $\chi$ is very small it is impossible to mix the polymers.

To see this in more detail, let us calculate $f_m$ for the case of polymers having equal number of monomers. Setting $N_A = N_B = N$, and writing $\phi_A = \phi, \phi_B = 1 - \phi$, we see that (2.85) becomes

$$f_m = \frac{1}{N}[\phi \ln \phi + (1 - \phi) \ln(1 - \phi)] + \chi\phi(1 - \phi). \tag{2.86}$$

Equation (2.86) is symmetric about $\phi = 1/2$, and so if the curve $f_m(\phi)$ has minima, the line joining them is also a common tangent to the curve at those points. Therefore, the coexistence curve of the phase diagram is determined by the positions of the minima. The positions of the minima are determined from $\partial f_m/\partial\phi = 0$, giving us

$$\frac{1}{1-2\phi}\ln\frac{\phi}{1-\phi} = -N\chi. \tag{2.87}$$

If $N\chi \gg 1$, the solutions to this are

$$\phi = \exp(-N\chi)\ ,\ 1-\phi = \exp(-N\chi). \tag{2.88}$$

Therefore there are two phases, consisting of almost pure A or pure B. Further, the critical point is found by solving $\partial^2 f_m/\partial\phi^2 = \partial^3 f_m/\partial\phi^3 = 0$, giving us

$$\phi_c = \frac{1}{2}\ ,\ \chi_c = \frac{2}{N}. \tag{2.89}$$

Thus, for polymers A and B to mix we must have $\chi < 2/N$, which implies that the interaction energy $\Delta\epsilon$ must be of order $1/N$. There are very few combinations of polymers which satisfy this, and this is why it is usually impossible to mix different polymers.

If $N_A$ and $N_B$ are not equal it is generally difficult to express the co-existence curve in a simple form. However, it is not hard to determine the critical point, given by the following equations:

$$\phi_{Ac} = \left(1 + \sqrt{\frac{N_A}{N_B}}\right)^{-1}\ ,\ \chi_c = \frac{1}{2}\left(\frac{1}{\sqrt{N_A}} + \frac{1}{\sqrt{N_B}}\right)^2. \tag{2.90}$$

Thus, we see that the critical concentration is weighted towards the low molecular weight component, and $\chi_c$ is determined mostly by the molecular weight of the small molecules.

### 2.3.2 Correlation function in miscible systems

Although there are not many, there are some examples of miscible polymer combinations. The density correlation function for such systems is given by (2.69). In particular, for mixtures with $N_A = N_B = N$, use of (1.38) for $g(q)$ gives the following:

$$S(q) = \frac{S(0)}{1 + q^2\xi^2}. \tag{2.91}$$

Here

$$S(0) = \frac{N\phi(1-\phi)}{1 - 2\chi N\phi(1-\phi)} \tag{2.92}$$

$$\xi^2 = \frac{Nb^2}{12[1 - 2\chi N\phi(1-\phi)]}. \tag{2.93}$$

If we are far enough removed from the critical point, the denominators of (2.92), (2.93) are of order 1, and we see that the scattering intensity is proportional to N, and the correlation length $\xi$ is proportional to $\sqrt{N}$.

## 2.4 Block copolymers

### 2.4.1 Block copolymers and microscopic phase separation

Block copolymers are made by joining polymers of two or more different types. As shown in Fig. 2.8, examples of the types of block copolymers that can be made from polymers of two types A, B include the A — B type, the A — B — A type and the grafted type.

Polymers of different types are generally not miscible, and so the A molecules and B molecules making up a block copolymer are usually immiscible. Thus, in a melt of a block copolymer, the A parts tend to cluster together as do the B parts, making a domain structure. Since the size of each domain cannot exceed the length of a stretched polymer chain, they are usually less than 1 micrometre. The appearance of these domains is called a microscopic phase separation.

The type of domain structure produced by a microscopic phase separation depends on the linear structure and chain length of the polymers. For A — B type block copolymers, the domain structure is determined by the ratio of the number of monomers in an A chain $N_A$ to the number in a B chain $N_B$. As shown in Fig. 2.9, if $N_A$ is smaller than $N_B$, the B chains form a continuous phase which contain spherical domains made of the A chains. If $N_A$ is increased, the domains containing the A chains change to a cylindrical shape. If the ratio of $N_A$ to $N_B$ is close to 1, there appears a laminar structure, with alternate layers of A and B. If the ratio of $N_A$ to $N_B$ is further increased, A chains now form the continuous phase with B chains in cylindrical or spherical domains.

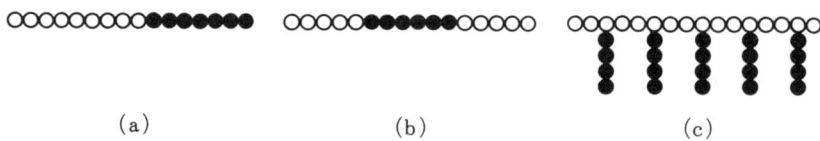

(a)             (b)             (c)

**Fig. 2.8** Examples of block copolymers. The white and black circles represent monomers A and B, respectively: (a) A—B block copolymer (a block of A monomers is connected to a block of B monomers); (b) A—B—A block copolymer; (c) grafted copolymer (chains of B monomers are connected ('grafted') to the chain of A monomers).

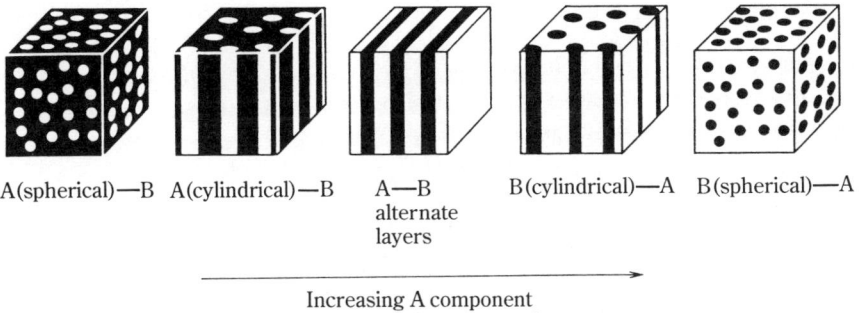

A(spherical)—B    A(cylindrical)—B    A—B    B(cylindrical)—A    B(spherical)—A
alternate
layers

Increasing A component
(decreasing B component)

**Fig. 2.9** The various phases formed by A—B block copolymers. The ratio $N_A/N_B$ is increasing from left to right.

### 2.4.2 Correlation function in the uniform phase

Since the theoretical treatment of a system with a domain structure is very difficult, we will first of all assume that the A chains and B chains are uniformly mixed and calculate the concentration fluctuations for each component. Assume that we have an A—B block copolymer with $N$ segments and let $f$ be the fraction of segments of type A. The number of monomers of each type can be written as follows.

$$N_A = Nf, \; N_B = N(1 - f). \tag{2.94}$$

As in the previous section, let us consider external potentials $u_A(r), u_B(r)$ which act on A and B, and calculate the change in the spatial distribution of the segments. The only difference to the calculation of the previous section is that here $S_{AB}^{(0)}(r) = S_{BA}^{(0)}(r)$ is not zero, since the A chains and B chains are joined together. Therefore, the equations corresponding to (2.63), (2.64) are

$$\overline{\phi_A} = -\beta \left[ S_{AA}^{(0)} u_A^{(\text{eff})} + S_{AB}^{(0)} u_B^{(\text{eff})} \right]$$

$$\overline{\phi_B} = -\beta \left[ S_{AB}^{(0)} u_A^{(\text{eff})} + S_{BB}^{(0)} u_B^{(\text{eff})} \right]. \tag{2.95}$$

Here

$$u_A^{(\text{eff})} = u_A - z\left( \epsilon_{AA} \overline{\phi_A} + \epsilon_{AB} \overline{\phi_B} \right) + V$$

$$u_B^{(\text{eff})} = u_B - z\left( \epsilon_{AB} \overline{\phi_A} + \epsilon_{BB} \overline{\phi_B} \right) + V. \tag{2.96}$$

Combining this with $\overline{\phi_A} + \overline{\phi_B} = 0$, and solving, gives us

$$\overline{\phi_A} = -\beta \left[ \frac{S_{AA}^{(0)} + S_{BB}^{(0)} + 2S_{AB}^{(0)}}{S_{AA}^{(0)} S_{BB}^{(0)} - (S_{AB}^{(0)})^2} \quad 2\chi \right]^{-1} (u_A - u_B). \tag{2.97}$$

$S_{AA}^{(0)}$ is calculated for a chain with Gaussian distribution as follows:

$$S_{AA}^{(0)} = \frac{1}{N} \int_0^{N_A} dn \int_0^{N_A} dm \exp\left(-\frac{b^2}{6}q^2 \,|\, n-m \,|\right) \equiv Nh(f,x) \tag{2.98}$$

Here $x, h(f,x)$ are defined as

$$x = q^2 R_g^2 = \frac{Nb^2}{6}q^2 \tag{2.99}$$

$$h(f,x) = \frac{2}{x^2} \left[fx + e^{-fx} - 1\right]. \tag{2.100}$$

A similar calculation also gives

$$S_{BB}^{(0)} = Nh(1-f,x). \tag{2.101}$$

Further

$$S_{AB}^{(0)} = \frac{1}{N} \int_0^{N_A} dn \int_{N_A}^{N} dm \exp\left(-\frac{b^2}{6}q^2 \,|\, n-m \,|\right)$$

$$= \frac{N}{2}[h(1,x) - h(f,x) - h(1-f,x)]. \tag{2.102}$$

Substituting these results into (2.97), the following equation is eventually obtained for the correlation function:

$$S(q) = \frac{N}{F(x) - 2\chi N} \tag{2.103}$$

where

$$F(x) = \frac{h(1,x)}{h(f,x)h(1-f,x) - \frac{1}{4}[h(1,x) - h(f,x) - h(1-f,x)]^2}. \tag{2.104}$$

If $x \gg 1$, in other words $qR_g \gg 1$, the correlation function $S(q)$ takes the following form:

$$S(q) = \frac{12f(1-f)}{q^2 b^2}. \tag{2.105}$$

This result is due to the fact that locally the polymers are connected, and does not reflect the block copolymer structure. In fact (2.105) can be obtained from (2.91) by taking the limit of $q\xi \gg 1$, and replacing $\phi$ by $f$.

On the other hand, if $qR_g \ll 1$, eqn (2.103) takes the following form:

$$S(q) = \frac{2}{3}Nf^2(1-f)^2 q^2 R_g^2. \tag{2.106}$$

So we see that for $qR_g \to 0$, we have also $S(q) \to 0$, which is a feature of block copolymers. A block copolymer is made of A chains linked to B chains, and so if we consider a region much larger than the maximum length of the chains, the number of A segments contained in that region is constant, and thus the concentration fluctuation is zero. This is why $S(q) \to 0$ as $q \to 0$.

As can be seen from (2.105), (2.106), the function $S(q)$ is an increasing function of $q$ for small $q$, and for large $q$ it is a decreasing function. Thus, $S(q)$ must have a maximum for an intermediate value of $q$.

Fig. 2.10 shows the result of numerical calculations of (2.103). For $qR_g \ll 1$, or $qR_g \gg 1$, the behaviour of $S(q)$ is independent of the miscibility parameter $\chi$, but in the intermediate region the behaviour depends strongly on $\chi$. In particular, as $\chi$ increases the maximum value of $S(q)$ diverges. This corresponds to the spinodal point of a two-component polymer system, and can be interpreted as the point when microscopic phase separation begins. As can be seen from (2.103), the critical value $\chi_c$ in this case is proportional to $1/N$ which is the same as two-component polymer systems (see eqn (2.89)). Therefore, under normal conditions $\chi > \chi_c$, and so most block copolymer systems are microscopically phase separated.

### 2.4.3 Domain size

Here we will estimate the domain size in an A-B block copolymer system. In the case of the lamellar structure formed when $f = 1/2$, the following simple argument can be made. Let us consider the free energy of a lamellar system

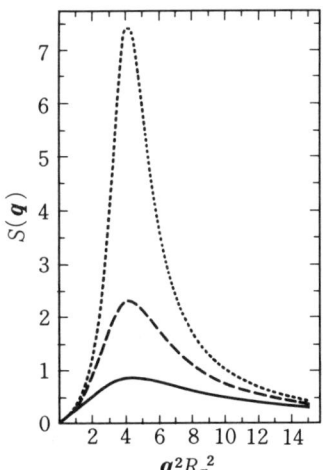

**Fig. 2.10** The structure factor of a block copolymer, with $f = 0.25$. The value of $\chi N$ for each curve is (from the top) 17.5, 16, 12.5. Here $S(q)$ diverges for $\chi N = 18.2$ (Leibler, L. (1980). *Macromolecules*, **13**, 1602.)

with layer thickness $D$ located in an arbitrary cube of edge length L. Writing $\sigma$ for the surface energy per unit area, the overall surface energy becomes

$$F_s \simeq 2\sigma \frac{L}{D} L^2. \tag{2.107}$$

So $F_s$ decreases as $D$ increases. However, as $D$ increases each polymer becomes stretched, causing the free energy to increase. For a polymer with end-to-end distance $D$, the free energy is of the order $k_B T D^2 / N b^2$. In the system there are $L^3 / N v_c$ polymers, and so the overall free energy due to the stretching of the polymers is

$$F_{el} \simeq k_B T \frac{D^2}{N b^2} \frac{L^3}{N v_c}. \tag{2.108}$$

The value of $D$ which minimizes the sum of $F_s$ and $F_{el}$ is found to be

$$D \simeq \left( \frac{\sigma N^2 b^2 v_c}{k_B T} \right)^{1/3} \propto N^{2/3}. \tag{2.109}$$

The result $D \propto N^{2/3}$ agrees approximately with experimental results.

# 3
# Polymer gels

A polymer gel is a three-dimensional network of polymer chains joined together at a number of connection sites, as shown in Fig. 3.1. The connections may be due to covalent chemical bonds or physical interactions such as hydrogen bonds or electrostatic forces. The former are called chemical gels, the latter physical gels.

There are basically two ways to make a chemical gel. One way is to add a cross linking agent to a system of polymer chains, causing them to form a network. Another way is to mix trivalent (or even higher valency) segments like $-B\!\!<$ with the usual divalent segments $-A-$ during the polymerization reaction. With either method, during the initial stages of the reaction there is a wide range of polymer sizes in the system, but as the reaction proceeds there appears one giant polymer spanning the entire system. This stage is called gelation.

Rubber is a good example of a polymer gel. Rubber is a giant network molecule made by a cross linking reaction between polymers such as isoprene. Before gelation the system is in the liquid state and takes the shape of the container holding it, but after gelation it maintains a fixed shape. For example, rubber will deform under an applied force, but will return to its original shape after the force is removed. In this sense we say that rubber behaves as an elastic solid. However, if we use X-ray scattering to look at the

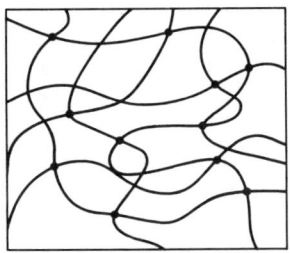

**Fig. 3.1** A polymer gel.

atomic structure at the nanometre scale, we see that the rubber state is almost indistinguishable from the liquid state. At the microscopic atomic level, rubber has a liquid-like or amorphous structure. It is one of the major features of polymeric materials that they show elastic behaviour whilst having a liquid-like structure.

Gelation is a complicated process involving irreversible chemical reactions. For example, if a connection is formed between two polymer chains in a solvent, the chains will crowd in around that region, and the cross linking reaction will be accelerated there. Because of this, if the cross linking reaction occurs slowly, a very non-uniform gel is formed. Thus we see that the structure of the gel greatly differs with the method of formation, and this is one of the reasons why the theoretical and experimental research of gels is so difficult. On the other hand, if the cross linking reaction proceeds quickly, it is possible to make a gel with a very uniform network. From now on, we will mainly restrict our discussion to gels of this type.

## 3.1 Elasticity of rubber

### 3.1.1 Polymer thermal motion and the elasticity of rubber

The elasticity of rubber is very different from that of crystalline solids. The elasticity modulus of rubber is $10^{-5}$ times smaller than that of steel, etc., and it can be stretched by several hundred % without breaking. Further, the elastic constant of rubber increases with temperature, and is approximately proportional to absolute temperature. These features can be nicely explained if we assume that the elasticity of rubber is due to the thermal motion of the polymer chains.

To see how the thermal motion of the polymers can lead to elasticity, let us consider the tension in a single ideal chain when we pull on its ends. Assume that the chain has $N$ segments and that one end is held at the origin, with the other end at $r$. Using the lattice model of the polymer of Section 1.3.1, the number of states the ideal chain can take under these circumstances is $W_0(r)$. Since the energy of the ideal chain is independent of $r$, the free energy can be written

$$A_{\text{chain}} = -k_B T \ln W_0(r) + \text{constant} = \frac{3k_B T}{2Nb^2} r^2 + \text{constant}. \quad (3.1)$$

Therefore, the tension pulling on the ends is

$$f = -\frac{\partial A_{\text{chain}}}{\partial r} = -\frac{3k_B T}{Nb^2} r. \quad (3.2)$$

Note that the reason for the tension $f$ is not the change in the energy but the change in the entropy. If the ends of a polymer are pulled, the number of allowable configurations that the polymer can take is reduced. The tension acts to return the chain to a state with more allowable configurations, that is, a state with higher entropy.

Actually, a similar situation occurs in an ideal gas, because if we confine the gas in a container we reduce the configurational entropy of the molecules, and this is what causes the gas to have pressure. Another way of saying this is that the thermal motion of the particles in the gas acts as a pressure to push the container walls apart. Similarly for polymers, the thermal motion of the segments acts to pull the ends of the polymer towards the centre.

Equation (3.2) means that the polymer behaves as a spring with spring constant $k = 3k_B T/Nb^2$. The size of this polymer spring is of the order of $\sqrt{N}b$ in the equilibrium state, and increases up to $Nb$ if the polymer is fully stretched. In other words, the material can be extended up to $\sqrt{N}$ times, and if $N$ is very large, this maximum extension also becomes large. This is why rubber can be stretched so much.

### 3.1.2 Elastic free energy of rubber

The fact that the elasticity of rubber derives from the thermal motion of the polymers was recognized by Kuhn in the early days of polymer science. Based on this idea, Kuhn was able to derive a simple but physically insightful theory of rubber elasticity. Let us now look at this classic theory.

Let us assume that a rubber sample undergoes a deformation, with a material element located at $R$ displaced to a new position $R'$. The tensor $E_{\alpha\beta} = \partial R'_\alpha/\partial R_\beta$ is called the deformation gradient tensor. A deformation where $E_{\alpha\beta}$ is constant is called a homogeneous deformation, and in this case we can write

$$R'_\alpha = E_{\alpha\beta} R_\beta. \tag{3.3}$$

To calculate the elastic free energy of the rubber due to the deformation, let us consider the free energy of the section of the polymer chain lying between two neighbouring connection sites (let us call this a partial chain). The free energy of a partial chain made of $N$ segments with end-to-end vector $r$ is $(3k_B T/2Nb^2)r^2$, and so the total free energy of the rubber can be written as follows:

$$A = n_c \int dr \int_0^\infty dN \Psi(r, N) \frac{3k_B T}{2Nb^2} r^2 + A_0(V, T). \tag{3.4}$$

Here, $n_c$ is the number of partial chains in the rubber, and $\Psi(r, N)$ is the probability of having a partial chain of $N$ segments with end-to-end vector $r$. To find $\Psi(r, N)$, it is necessary to know the distribution of the partial chains in the undeformed rubber, as well as how the connection sites move when the sample is deformed. These are difficult problems, and so Kuhn made the following drastic assumptions.

1. The distribution function of the end-to-end vector $r$ of the partial chain in the undeformed state is Gaussian, the same as the distribution function of a polymer in solution. Therefore, letting $\Phi_0(N)$ be the distribution function for $N$, we can write the distribution function for $r$, $N$ as follows:

$$\Psi_0(r, N) = \left(\frac{3}{2\pi Nb^2}\right)^{3/2} \exp\left(-\frac{3r^2}{2Nb^2}\right)\Phi_0(N). \tag{3.5}$$

Here

$$\int_0^\infty dN\Phi_0(N) = 1. \tag{3.6}$$

2. Under the deformation, the connection sites move affinely with the macroscopic deformation. In other words, if the end-to-end vector of the partial chain is initially $r$, after the deformation the vector changes to $r' = E \cdot r$. Therefore, the free energy becomes

$$A = n_c \int dr \int dN\Psi_0(r, N)\frac{3k_BT}{2Nb^2}(E \cdot r)^2 + A_0(V, T). \tag{3.7}$$

Substituting (3.5) into (3.7) and calculating gives

$$A = \frac{1}{2}n_c k_B T(E_{\alpha\beta})^2 + A_0(V, T). \tag{3.8}$$

### 3.1.3 Relationship between stress and strain

If we know the free energy change due to the deformation, the stress tensor $\sigma_{\alpha\beta}$ can be determined as follows. If a material under a stress $\sigma_{\alpha\beta}$ is subject to an infinitesimal strain $\delta\epsilon_{\alpha\beta}$, with a point at position vector $R$ shifting by an amount $\delta R_\alpha = \delta\epsilon_{\alpha\beta}R_\beta$, then the work done on the system is given by $\delta W = V\sigma_{\alpha\beta}\delta\epsilon_{\alpha\beta}$ (here $V$ is the system volume). Under isothermal conditions, the work done equals the change in the free energy $\delta A$, which gives us

$$\delta A = V\sigma_{\alpha\beta}\delta\epsilon_{\beta\alpha}. \tag{3.9}$$

Under the infinitesimal strain $\delta\epsilon_{\alpha\beta}$, the deformation gradient tensor and the volume will change as follows:

$$\delta E_{\alpha\beta} = \delta\epsilon_{\alpha\mu}E_{\mu\beta} \ , \ \delta V = V\delta\epsilon_{\mu\mu}. \tag{3.10}$$

From (3.8) we obtain

$$\delta A = n_c k_B T\delta\epsilon_{\alpha\mu}E_{\mu\beta}E_{\alpha\beta} + \left(\frac{\partial A_0}{\partial V}\right)_T V\delta\epsilon_{\mu\mu}. \tag{3.11}$$

Comparing (3.9) with (3.11) gives us

$$\sigma_{\alpha\beta} = \nu_c k_B T E_{\alpha\mu} E_{\beta\mu} - P\delta_{\alpha\beta}. \tag{3.12}$$

Here $\nu_c = n_c/V$ is the number of partial chains per unit volume, and $P = -(\partial A_0/\partial V)_T$.

As an example, let us consider the shearing deformation of Fig. 3.2(a). In this case

$$x' = x + \gamma y \ , \ y' = y \ , \ z' = z, \tag{3.13}$$

where $\gamma$ is called the shear strain. From (3.13) we find that

$$(E_{\alpha\beta}) = \begin{pmatrix} 1 & \gamma & 0 \\ 0 & 1 & 0 \\ 0 & 0 & 1 \end{pmatrix}. \tag{3.14}$$

Therefore, the stress tensor becomes

$$\sigma_{xy} = \nu_c k_B T \gamma \tag{3.15}$$
$$\sigma_{xx} = \nu_c k_B T(\gamma^2 + 1) - P \tag{3.16}$$
$$\sigma_{yy} = \sigma_{zz} = \nu_c k_B T - P. \tag{3.17}$$

The shear elastic modulus $G$ is defined as follows:

$$G = \lim_{\gamma \to 0} \frac{1}{\gamma} \sigma_{xy}. \tag{3.18}$$

Equation (3.15) therefore gives us

$$G = \nu_c k_B T. \tag{3.19}$$

In other words, the shear elastic modulus is proportional to the number density of the partial chains $\nu_c$.

Although the pressure $P$ in (3.12) is a function of $V$ and $T$, rubber is similar to liquids in that its volume is almost constant. So let us assume that rubber is incompressible and has a constant volume, which means that

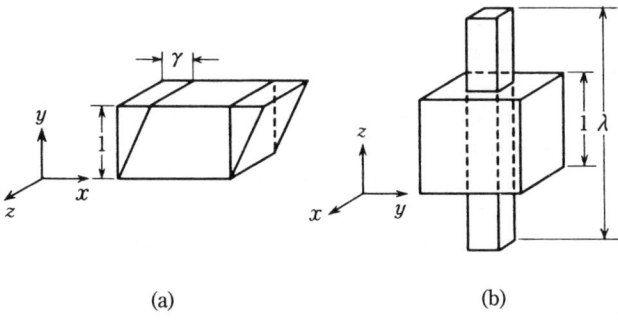

(a)                                  (b)

**Fig. 3.2** (a) Shearing deformation; (b) uniaxial extension.

$P$ will be determined by the external conditions (eg. if the sample is at atmospheric pressure, or at the bottom of the sea, etc.). Thus, in order to study the intrinsic properties of rubber it is convenient to consider material properties that do not depend on $P$, and the following material functions are often used:

$$N_1 = \sigma_{xx} - \sigma_{yy} \;,\; N_2 = \sigma_{yy} - \sigma_{zz}. \tag{3.20}$$

These are called, respectively, the first normal stress difference and second normal stress difference.

As another example, let us consider the uniaxial extension shown in Fig. 3.2(b). Assuming that the material is stretched by a factor $\lambda$ in the $z$ direction, the condition of constant volume requires that the dimensions in the $x$ and $y$ directions be reduced by a factor of $1/\sqrt{\lambda}$. Therefore we have

$$(E_{\alpha\beta}) = \begin{pmatrix} 1/\sqrt{\lambda} & 0 & 0 \\ 0 & 1/\sqrt{\lambda} & 0 \\ 0 & 0 & \lambda \end{pmatrix}, \tag{3.21}$$

and the stress becomes

$$\sigma_{xx} = \sigma_{yy} = \frac{\nu_c k_B T}{\lambda} - P \;,\; \sigma_{zz} = \nu_c k_B T \lambda^2 - P. \tag{3.22}$$

Since no force acts on the side walls of the specimen, we have $\sigma_{xx} = \sigma_{yy} = 0$. This means that $P = \nu_c k_B T/\lambda$, and so $\sigma_{zz}$ becomes

$$\sigma_{zz} = \nu_c k_B T (\lambda^2 - \frac{1}{\lambda}). \tag{3.23}$$

According to (3.23), the extension $\lambda$ can be made arbitrarily large if the applied stress is large enough. In reality, a chain made up of $N$ segments cannot be extended more than a length of $Nb$, and so (3.23) does not hold for very large extensions. To correct for this, it is necessary to introduce a model of the chain which takes into account the effects of finite extensibility.

However, the relationsip between stress and strain in (3.23) becomes invalid even before finite extension effects occur. In Kuhn's theory, the effects of fluctuations in the position of the connection sites and the interaction between the partial chains have been completely ignored. Thus, it is not surprising that there is a discrepancy between experiment and theory.

Despite these faults, the reason why Kuhn's theory is regarded as a classic work is that it pinpointed the entropy of the chains as the origin of rubber elasticity. In actual fact, the following law derived from Kuhn's theory is found to agree very well with real polymer systems.

## 3.2 The stress optical law

*3.2.1 Orientation of bonds of a stretched chain*

In Section 3.1.1 we saw that the reason for the elasticity of the polymer chain is the decrease in the entropy of the chain when it is stretched. Let us now look at this from a different angle. The entropy of a chain has its origins in the fact that the constituent elements, the bonds, have a directional degree of freedom. Writing the unit direction vector of a particular bond as $u$, and the distribution function for the direction as $f(u)$, the entropy of a chain made of $N$ bonds can be expressed as follows:

$$S_{\text{chain}} = -Nk_B \int du f(u) \ln f(u). \tag{3.24}$$

Here $\int du$ is the surface integral over the unit sphere $|\, u \,| = 1$. In the equilibrium state $u$ has an isotropic distribution, but if the polymer is stretched by forces acting on its ends the distribution of $u$ is biased in the extension direction. Because of this the entropy decreases, and this gives rise to the restoring force in the chain.

The degree of alignment of the bonds can be represented by the directional order parameter tensor, defined as follows:

$$Q_{\alpha\beta} \equiv \int du \left( u_\alpha u_\beta - \frac{1}{3}\delta_{\alpha\beta} \right) f(u). \tag{3.25}$$

If $f(u)$ is an isotropic distribution $Q_{\alpha\beta}$ becomes zero, and if there is a bias in the distribution of $u$ the tensor $Q_{\alpha\beta}$ is non-zero. Therefore we may expect that there is a relationship between the elasticity of the polymer and $Q_{\alpha\beta}$. Actually, as we will show later, $Q_{\alpha\beta}$ is proportional to the stress tensor $\sigma_{\alpha\beta}$:

$$Q_{\alpha\beta} = C_1 \left( \sigma_{\alpha\beta} - \frac{1}{3}\delta_{\alpha\beta}\sigma_{\mu\mu} \right). \tag{3.26}$$

Kuhn was the first to derive this relationship.

In order to prove (3.26), let us use as a model of the polymer the freely jointed chain shown in Fig. 3.3, which is a type of ideal chain. Assume that bonds of fixed length $b$ are free to rotate about their connection sites, with their direction independent of other bonds. Label the bonds $1, 2, ..., N$, and write $u_n$ for the unit direction vector of the $n$th bond. If the end-to-end vector of this chain is $r$, the following equation must hold:

$$\sum_{n=1}^{N} u_n = \frac{r}{b}. \tag{3.27}$$

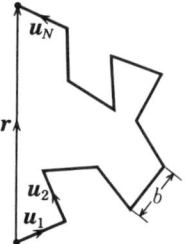

**Fig. 3.3** A freely jointed chain.

Note that under a condition such as (3.27), the distribution function $f(\boldsymbol{u})$ will not necessarily be isotropic. Let us now calculate $f(\boldsymbol{u})$. If $N$ is large, the condition of (3.27) can be re-expressed as

$$\int \mathrm{d}\boldsymbol{u} f(\boldsymbol{u})\boldsymbol{u} = \frac{\boldsymbol{r}}{Nb}. \tag{3.28}$$

On the other hand, the free energy of the freely-jointed chain is given by

$$A_{\text{chain}} = -TS_{\text{chain}} = Nk_{\text{B}}T \int \mathrm{d}\boldsymbol{u} f(\boldsymbol{u}) \ln f(\boldsymbol{u}). \tag{3.29}$$

Therefore, the equilibrium distribution $f(\boldsymbol{u})$ is found by minimizing (3.29) under the condition of (3.28). Taking into account condition (3.28) and the normalization condition, we need to minimize the following:

$$\tilde{A}_{\text{chain}} = Nk_{\text{B}}T\left[\int \mathrm{d}\boldsymbol{u} f \ln f - \boldsymbol{\lambda} \cdot \int \mathrm{d}\boldsymbol{u} f(\boldsymbol{u})\boldsymbol{u} - \mu \int \mathrm{d}\boldsymbol{u} f(\boldsymbol{u})\right]. \tag{3.30}$$

Taking the variation of (3.30) with respect to $f$ and setting this equal to zero yields the following:

$$f(\boldsymbol{u}) = \frac{1}{Z(\lambda)}\exp(\boldsymbol{\lambda} \cdot \boldsymbol{u}). \tag{3.31}$$

Here, the normalization constant $Z$ is given by

$$Z(\lambda) = \int \mathrm{d}\boldsymbol{u}\exp(\boldsymbol{\lambda} \cdot \boldsymbol{u}) = 4\pi\frac{\sinh \lambda}{\lambda}. \tag{3.32}$$

The constant $\boldsymbol{\lambda}$ is determined by (3.28). Substituting (3.31) into (3.28) gives

$$\frac{\partial}{\partial \boldsymbol{\lambda}}\ln Z(\lambda) = \frac{\boldsymbol{r}}{Nb}. \tag{3.33}$$

From this, we see that $\boldsymbol{\lambda}$ is parallel to $\boldsymbol{r}$, and its magnitude is given by the following equation:

$$\coth \lambda - \frac{1}{\lambda} = \frac{r}{Nb}. \tag{3.34}$$

So long as the polymer is not near its stretching limit, $r$ is of the order of $\sqrt{Nb}$, and so we can assume that $r/Nb \ll 1$. Therefore, the solution of (3.34) can be obtained by an expansion with respect to $r/Nb$ :

$$\lambda = \frac{3r}{Nb} + \frac{9}{5}\left(\frac{r}{Nb}\right)^3 + \cdots \tag{3.35}$$

Substituting (3.31) into (3.29), and using (3.32) and (3.35), gives us

$$A_{\text{chain}} = Nk_B T\left(\frac{\lambda \cdot r}{Nb} - \ln Z\right)$$

$$= \frac{3k_B T}{2Nb^2}r^2 + \frac{9N}{20}k_B T\left(\frac{r}{Nb}\right)^4 + \cdots \tag{3.36}$$

The first term of (3.36) agrees with (3.1).

Let us now calculate the directional order parameter tensor for the distribution function (3.31). First of all we define the following average using the isotropic distribution of $u$:

$$\langle \ldots \rangle_0 = \frac{1}{4\pi}\int d u \ldots. \tag{3.37}$$

Now, substituting (3.31) into (3.25) gives us

$$Q_{\alpha\beta} = \frac{\langle (u_\alpha u_\beta - \frac{1}{3}\delta_{\alpha\beta})\exp(\lambda \cdot u)\rangle_0}{\langle \exp(\lambda \cdot u)\rangle_0}. \tag{3.38}$$

Expanding $\exp(\lambda \cdot u)$ with respect to $\lambda$, and using the following identities

$$\langle u_\alpha \rangle_0 = 0$$

$$\langle u_\alpha u_\beta \rangle_0 = \frac{1}{3}\delta_{\alpha\beta}$$

$$\langle u_\alpha u_\beta u_\mu u_\nu \rangle_0 = \frac{1}{15}(\delta_{\alpha\beta}\delta_{\mu\nu} + \delta_{\alpha\mu}\delta_{\beta\nu} + \delta_{\alpha\nu}\delta_{\beta\mu}) \tag{3.39}$$

$$\langle \text{product of an odd number of } u_\alpha \ldots \rangle_0 = 0 \tag{3.40}$$

gives us

$$Q_{\alpha\beta}(r) = \frac{1}{15}\left(\lambda_\alpha \lambda_\beta - \frac{\lambda^2}{3}\delta_{\alpha\beta}\right) + O(\lambda^4). \tag{3.41}$$

Here, if we substitute (3.35) we obtain

$$Q_{\alpha\beta}(r) = \frac{3}{5N^2b^2}\left(r_\alpha r_\beta - \frac{r^2}{3}\delta_{\alpha\beta}\right). \tag{3.42}$$

Note that $Q_{\alpha\beta}(r) \simeq 1/N$ for $r \simeq \sqrt{Nb}$.

### 3.2.2 Stress tensor and orientational order parameter tensor

The above discussion is only concerned with a single partial chain in the rubber, and so we must take an average over many partial chains to calculate the overall material properties. If we choose at random a segment in the rubber, then the probability that it is a member of a partial chain made of $N$ segments with end-to-end vector $r$ is proportional to $N\Psi(r, N)$. Therefore, if we average (3.42) with respect to this weight, we have

$$Q_{\alpha\beta} = \frac{\int dr \int dN \frac{3}{5Nb^2}\left(r_\alpha r_\beta - \frac{r^2}{3}\delta_{\alpha\beta}\right)\Psi(r, N)}{\int dr \int dN N\Psi(r, N)}. \tag{3.43}$$

On the other hand, the free energy of the rubber is given by (3.4), and under an infinitesimal deformation $\delta\epsilon_{\alpha\beta}$ the position vector $r_\alpha$ changes by $\delta r_\alpha = \delta\epsilon_{\alpha\beta}r_\beta$, giving us

$$\delta A = n_c \int dr \int dN\Psi(r, N)\frac{3k_B T}{Nb^2}r_\alpha r_\beta \delta\epsilon_{\alpha\beta} + \frac{\partial A_0}{\partial V}V\delta\epsilon_{\mu\mu}. \tag{3.44}$$

The stress is calculated from this and (3.9) and is found to be

$$\sigma_{\alpha\beta} = \nu_c \int dr \int dN\Psi(r, N)\frac{3k_B T}{Nb^2}r_\alpha r_\beta - P\delta_{\alpha\beta}. \tag{3.45}$$

Comparing (3.43) and (3.45), we obtain (3.26), thus achieving our goal.

If the segments are aligned so that $Q_{\alpha\beta}$ is no longer 0 the material properties will become anisotropic. As an example, let us consider the dielectric constant tensor $\varepsilon_{\alpha\beta}$, which is usually isotropic in the equilibrium state, but becomes anisotropic if the rubber is stretched. If $Q_{\alpha\beta}$ is small, we can assume that the anisotropic part of $\varepsilon_{\alpha\beta}$ is proportional to $Q_{\alpha\beta}$:

$$Q_{\alpha\beta} = C_2\left(\varepsilon_{\alpha\beta} - \frac{\delta_{\alpha\beta}}{3}\varepsilon_{\mu\mu}\right). \tag{3.46}$$

From (3.26) and (3.46) we obtain

$$\varepsilon_{\alpha\beta} - \frac{\delta_{\alpha\beta}}{3}\varepsilon_{\mu\mu} = C_3\left(\sigma_{\alpha\beta} - \frac{\delta_{\alpha\beta}}{3}\sigma_{\mu\mu}\right). \tag{3.47}$$

In particular, (3.47) tells us that the optical anisotropy (birefringence) of an isotropic material under an applied stress is proportional to the stress. This relationship is called the stress optical law, and holds not only for rubber, but also for polymer solutions and polymer melts.

The stress optical law also holds for amorphous solids such as glass, but there is a fundamental difference in the case of rubber. The stress optical law for usual solids holds only for very small deformations, and can be derived from symmetry considerations if we assume that the stress and dielectric constant tensors are linear functions of the strain.

However, in experiments with rubber materials, it has been confirmed that the stress optical law holds even if there is a non-linear relationship between stress and strain. Furthermore, the stress optical law holds even if there are relaxation effects between shear and strain such as in a polymer solution.

As an example, consider the experimental result in Fig. 3.4, which supports the stress optical law. Here the polymer solution is under a constant shear rate $\dot{\gamma} = \mathrm{d}\gamma/\mathrm{d}t$, and the time variation of shear stress $\sigma_{xy}$ and the first normal stress difference $N_1 = \sigma_{xx} - \sigma_{yy}$ have been plotted (in Fig. 3.4 these quantities have been divided by $\dot{\gamma}$ and $\dot{\gamma}^2$, respectively). The open circles are the measured values of the stress, and the line is the curve predicted from the birefringence using the stress optical law. The fact that the behaviour of the stress depends on the shear rate indicates that there are non-linear relaxation effects present, but clearly there is good agreement between the open circles and the predicted curve. We can see that even though there is a complex relationship between the deformation and the stress, there is still a simple linear relationship between the stress and birefringence, which is due to the fact that the stress is directly related to the anisotropy of the segment orientational distribution. Thus the stress

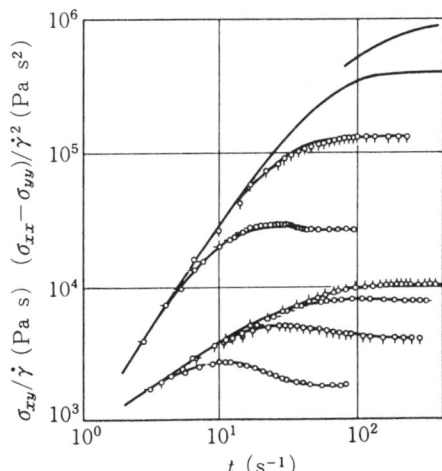

**Fig. 3.4** The time evolution of stress in a solution of polystyrene in alcohol after the commencement of shearing flow. The vertical axis represents the shear stress $\sigma_{xy}$ and the first normal stress difference $(\sigma_{xx} - \sigma_{yy})$, divided by $\dot{\gamma}$ and $\dot{\gamma}^2$, respectively. The circles are the measured values of the stress, and the lines are obtained from birefringence measurements using the stress optical law. (Takahashi, M., Masuda, T., Bessho, N., and Osaki, K., (1980). *J. Rheology*, **24**, 517, Fig. 1.)

optical law provides an important means for investigating the elastic nature of rubber materials.[1]

## 3.3 Interactions between partial chains

The theory presented in Section 3.2 contains a very big simplifying assumption. We assumed that the partial chains making up the rubber are completely free to move about, and experience no effects from the other chains other than those at the attached ends. If we think about the actual state of matter in rubber, it is clear that this model is very unrealistic. The polymer molecules in rubber are packed in very tightly, and the interactions between the molecules should be similar to those in the liquid state. In this section let us think about the effect of these interactions.

### 3.3.1 Excluded volume interactions

Let us first consider the effect of excluded volume interactions between the segments. As we saw in Chapter 2, excluded volume effects are extremely important when calculating the absolute value of the entropy of the chain. However, in the case of rubber elasticity, it is the difference in entropy between the extended and equilibrium states that is important. For this entropy difference, it is thought that the effects of excluded volume interactions are small. There are two reasons we can say this. The first reason is that if we use the lattice model and mean field approximation to calculate the entropy difference, as in Chapter 2, we will find that excluded volume interactions have no effects. In the Flory–Huggins theory for example, the effects of excluded volume interactions only appear through the volume fraction of the polymer $\phi$, and so if the material is deformed under constant volume conditions, excluded volume will have no effect. The second reason is that in the polymer melt state, it has been confirmed both theoretically and experimentally that each polymer molecule behaves as an ideal chain, and it is natural to expect a similar situation in the rubber state. In other words, the absolute value of the entropy of the partial chain will be changed due to excluded volume interactions, but the difference in entropy induced when the chain is stretched will not be altered. (This is not to say that there exists a rigorous proof of this assertion.)

---

[1]In the case of polymer melts there are no physical links between the polymers, and so the theoretical explanation for the stress optical law presented above is strictly not applicable. However, as we shall see later in Section 5.3, if we divide the polymer in the melt into partial chains of appropriate length and use the arguments in this section, we can derive the stress optical law.

*3.3.2 Nematic interactions*

In general, in liquids consisting of rod-like molecules, there is an interaction between neighbouring molecules tending to align them. If this interaction is strong, the system undergoes a phase transition to the nematic liquid crystal state, and so the interaction is called the nematic interaction.

The segments of a polymer have anisotropy similar to rod-like molecules, and so it is natural to expect that there will be nematic interactions between them. In actual fact, it has been observed that if rubber is stretched, not only the polymer segments but also the small molecules trapped between them are aligned in the stretching direction, which shows that there is a nematic interaction between the polymer segments and the small molecules. Here we will only consider the effect of nematic interactions between the polymer segments.

As a model of the polymer, we will use the freely jointed chain, and consider the interaction between two segments separated a distance $r$. Let $u$, $u'$ be the orientations of the two segments. Since the segments have axial symmetry, the interaction energy must be invariant under the transformation $u \rightarrow -u$. Also, the condition that energy is lowest when $u$ and $u'$ are parallel (or anti-parallel) means that the potential energy between the two segments must be a decreasing function of $(u \cdot u')^2$. The simplest potential satisfying these conditions is

$$U_{nem}(r, u, u') = -\tilde{E}_{nem}(r)\left[(u \cdot u')^2 - \frac{1}{3}\right]. \tag{3.48}$$

Here $\tilde{E}_{nem}(r)$ is a positive-valued function, and the constant $1/3$ has been added for mathematical convenience.

Assuming that all segments in the rubber are interacting through the potential (3.48), let us investigate the effects of this through the mean field approximation. The term in the square brackets in (3.48) can be written as $(u_\alpha u_\beta - \delta_{\alpha\beta}/3)(u'_\alpha u'_\beta - \delta_{\alpha\beta}/3)$. Taking the average of this over $u'$ and $r$, the mean molecular field $\bar{U}_{nem}$ experienced by a segment pointing in the $u$ direction can be calculated:

$$\bar{U}_{nem} = -E_{nem}Q_{\alpha\beta}\left(u_\alpha u_\beta - \frac{1}{3}\delta_{\alpha\beta}\right). \tag{3.49}$$

Here $Q_{\alpha\beta} = \langle u_\alpha u_\beta - \delta_{\alpha\beta}/3 \rangle$ is the segment order parameter tensor introduced in the previous section, and $E_{nem}$ is the average of $\tilde{E}_{nem}(r)$ over $r$. This $E_{nem}$ indicates the strength of the nematic interactions.

Repeating the arguments of Section 3.2.1 for the potential (3.49), we see that the orientational distribution function for the segments is the distribution (3.31) with the inclusion of the molecular field contribution $\exp(-\bar{U}_{\text{nem}}/k_B T)$ as follows:

$$f(u) = \frac{1}{Z(\lambda)} \exp\left[\lambda_\alpha u_\alpha + \epsilon Q_{\alpha\beta}\left(u_\alpha u_\beta - \frac{1}{3}\delta_{\alpha\beta}\right)\right]. \qquad (3.50)$$

Here $\epsilon = E_{\text{nem}}/k_B T$ and $\lambda$ is a parameter determined from (3.28) as before. Substituting (3.50) into (3.28) and expanding with respect to $\lambda_\alpha$, $Q_{\alpha\beta}$ gives

$$\frac{r_\alpha}{Nb} = \frac{\lambda_\alpha}{3} + \frac{2\epsilon}{15} Q_{\alpha\beta}\lambda_\beta. \qquad (3.51)$$

As we noted after (3.42), we have $Q_{\alpha\beta} \sim 1/N$ and so the relationship between $\lambda_\alpha$ and $r_\alpha$ can be written as follows.

$$\lambda_\alpha = \frac{3r_\alpha}{Nb}\left[1 + O\left(\frac{\epsilon}{N}\right)\right]. \qquad (3.52)$$

Furthermore, the free energy can be calculated in a way similar to (3.36):

$$A_{\text{chain}} = \frac{3k_B T}{2Nb^2} r^2 \left[1 + O\left(\frac{\epsilon}{N}\right)\right]. \qquad (3.53)$$

Therefore, we see that the effect of nematic interactions on rubber elasticity is of the order of $\epsilon/N$, and so can be neglected if $N \gg 1$.

On the other hand, if we calculate $Q_{\alpha\beta}(r)$ from (3.50) we get

$$
\begin{aligned}
Q_{\alpha\beta}(r) &= \frac{1}{15}\left(\lambda_\alpha\lambda_\beta - \frac{\lambda^2}{3}\delta_{\alpha\beta}\right) + \frac{2\epsilon}{15}Q_{\alpha\beta} \\
&= \frac{3}{5N^2 b^2}\left(r_\alpha r_\beta - \frac{r^2}{3}\delta_{\alpha\beta}\right) + \frac{2}{15}\epsilon Q_{\alpha\beta}. \qquad (3.54)
\end{aligned}
$$

If we multiply this by the weighting function $N\Psi(r, N)$ and take the average, we find

$$Q_{\alpha\beta} = \left\langle \frac{3}{5N^2 b^2}\left(r_\alpha r_\beta - \frac{r^2}{3}\delta_{\alpha\beta}\right)\right\rangle + \frac{2}{15}\epsilon Q_{\alpha\beta}. \qquad (3.55)$$

In other words,

$$Q_{\alpha\beta} = \frac{1}{1 - 2\epsilon/15}\left\langle \frac{3}{5N^2 b^2}\left(r_\alpha r_\beta - \frac{r^2}{3}\delta_{\alpha\beta}\right)\right\rangle. \qquad (3.56)$$

Comparing this with (3.43), we see that the stress optical law coefficient $C_3^{\text{nem}}$ in a system with nematic interactions is related to the coefficient $C_3$, in the absence of interactions, as follows:

$$C_3^{\text{nem}} = \frac{C_3}{1 - 2\epsilon/15}. \qquad (3.57)$$

Thus we see that nematic interactions only lead to a change in the coefficient in the stress optical law. However, in systems where nematic interactions are strong, such as rubber made up of liquid crystalline monomers, nematic interactions may give rise to some peculiar properties.

### 3.3.3 Entanglement interactions

Entanglement interactions are based on the restriction that the polymer chains cannot cut through each other. For example, let us consider the two cases (a) and (b) in Fig. 3.5. The partial chains shown in (a) can never take the configuration shown in (b). This is because polymers are one-dimensional continuous chains, and cannot cut across each other like ghosts or phantoms. The interaction based on this restriction is called the entanglement interaction. The models we have been using up to now have ignored this restriction, and so are often called 'phantom chain' models.

It is extremely difficult to treat entanglement interactions rigorously. Entanglement interactions exist even for ideal chains without excluded volume effects, and cannot be expressed through a potential. For this reason, entanglement effects had long been one of the most difficult problems confronting polymer theorists. However, with the introduction of the tube concept by Edwards and de Gennes, theoretical treatment became possible, albeit based on a simplifying model. Here we will follow the arguments of Edwards and estimate the effect of entanglements on rubber elasticity.

Let us consider a rubber made from extremely long polymers connected by a small amount of cross-linking agent. The number of allowable states of a partial chain in this rubber is very much smaller than if it were in free space. Fig. 3.6(a) shows schematically a snapshot of the partial chain placed on the plane of the page, with the ends A and B fixed. The dots

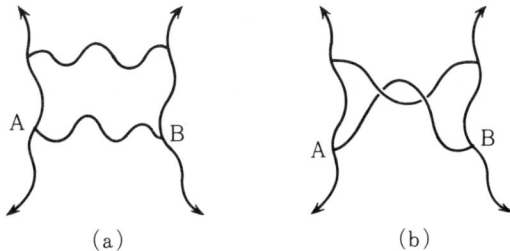

(a)                              (b)

**Fig. 3.5** Entanglement effects between a partial chain AB and the gel network. The chains with arrows are connected to the infinite network, and are assumed to be fixed.

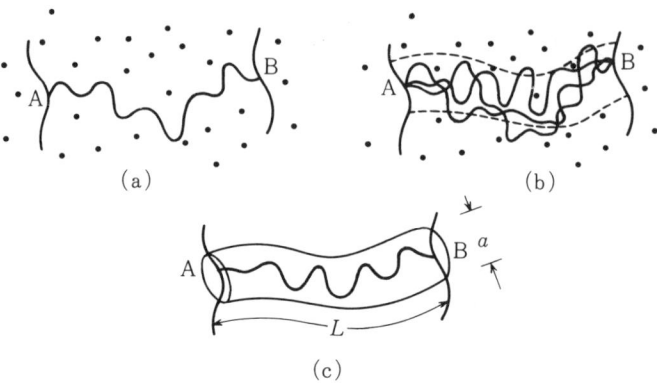

**Fig. 3.6** The tube model.

in the figure are the cross-sections of other polymers that cut through the plane, and we assume that the partial chain cannot cut across these dots. For simplicity we will assume that these dots do not move. The problem is now to calculate the number of allowable states of the partial chain in Fig. 3.6(a).

Even with these simplifications the problem is still mathematically very difficult, and so Edwards reasoned as follows. For most of the time, the partial chain in Fig. 3.6(a) is lying in the tubular region indicated in Fig. 3.6(b) with a dotted line. So instead of the network, let us consider only the tube as shown in Fig. 3.6(c), and assume that the partial chain is trapped inside. The diameter of the tube is of the same order as the spacing of the network $a$. This representation is called the tube model.

Let us now calculate the entropy of the polymer based on the tube model. Assume that the polymer is trapped in a tube of diameter $a$, with the ends of the chain affixed to two points A and B separated by a distance $L$ measured along the tube. In order to calculate the number of allowable states of this polymer, let us consider all the random walks which start at A and reach B after $N$ steps without going outside the tube. We will find it convenient to consider separately the component of the random walk parallel to the axis of the tube and the component perpendicular to the tube axis. These two components are independent, and so the total number of walks $W$ can be written as follows.

$$W = W_1(L, N)W_2(a, N). \tag{3.58}$$

Here $W_1(L, N)$ is the total number of walks in the direction of the tube axis, and $W_2(a, N)$ is the number of walks in the plane perpendicular to the tube. $W_1(L, N)$ is equal to the number of random walks in one dimension and so we have

$$W_1(L, N) = z_1^N \left( \frac{3}{2\pi Nb^2} \right)^{3/2} \exp\left( -\frac{3L^2}{2Nb^2} \right). \tag{3.59}$$

Further, $W_2(a, N)$ is the number of random walks in a circle of diameter $a$. If $N$ is large, this quantity is given as follows:

$$W_2(a, N) = z_2(a)^N. \tag{3.60}$$

Here $z_2(a)$ depends on the tube diameter.

Now, in order to calculate the change in entropy when the rubber is deformed, we have to know how $a$ and $L$ change under the deformation. Here for simplicity we will assume that the tube radius $a$ is unchanged and that the tube length $L$ is deformed affinely with the applied macroscopic deformation. Then the change in free energy due to the stretching is

$$A'_{\text{chain}} - A_{\text{chain}} = k_B T(\ln W' - \ln W) = \frac{3k_B T}{2Nb^2}(L'^2 - L^2). \tag{3.61}$$

Further, in order to calculate $L$ and $L'$, let us assume that in the undeformed state the central axis of the tube is a linkage of $Z$ small elements of

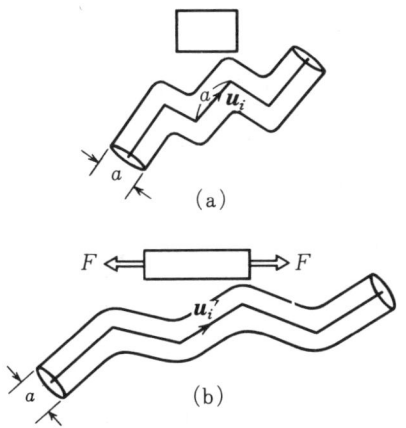

(a)

(b)

**Fig. 3.7** When rubber undergoes an extension, we make the following assumptions regarding the change in shape of the molecular tube. (a) In the undeformed state, the central axis of the tube is assumed to be a linkage of small elements of length $a$, oriented randomly. (b) When the rubber is stretched, the central axis of the tube deforms affinely with the macroscopic deformation, but the diameter of the tube remains $a$, as in the undeformed state.

length $a$ of random orientation, as shown in Fig. 3.7. If in the undeformed state an element was pointing in the direction of the unit vector $\boldsymbol{u}_i$, then after the deformation it will point in the direction $E \cdot \boldsymbol{u}_i$ and will have a length of $a \mid E \cdot \boldsymbol{u}_i \mid$. Thus $L'$ can be written as follows:

$$L' = \sum_{i=1}^{Z} a \mid E \cdot \boldsymbol{u}_i \mid . \tag{3.62}$$

Since $\boldsymbol{u}_i$ is a vector with an isotropic distribution, if $Z$ is large (3.62) becomes the following, where $\langle ... \rangle_0$ is the average with respect to the isotropic distribution:

$$L' = Za\langle \mid E \cdot \boldsymbol{u} \mid \rangle_0. \tag{3.63}$$

Therefore (3.61) becomes

$$A'_{\text{chain}} - A_{\text{chain}} = \frac{3k_{\text{B}}T}{2Nb^2} Z^2 a^2 \left( \langle \mid E \cdot \boldsymbol{u} \mid \rangle_0^2 - 1 \right). \tag{3.64}$$

Thus, writing the number of partial chains as $n_{\text{c}}$, the elastic free energy becomes

$$A = \frac{3}{2} n_{\text{c}} k_{\text{B}} T \frac{Z^2 a^2}{Nb^2} \langle \mid E \cdot \boldsymbol{u} \mid \rangle_0^2. \tag{3.65}$$

It is instructive to compare (3.65) with the result obtained from Kuhn's theory. Using the result $(E_{\alpha\beta})^2 = 3\langle \mid E \cdot \boldsymbol{u} \mid^2 \rangle_0$, (3.8) can be written as

$$A = \frac{3}{2} n_{\text{c}} k_{\text{B}} T \langle \mid E \cdot \boldsymbol{u} \mid^2 \rangle_0. \tag{3.66}$$

Comparing (3.65) with (3.66), we see that the coefficient in the former is $Z^2 a^2 / Nb^2$ times larger than that in the latter. Since $Z$ is proportional to $N$, this factor increases proportionally with $N$. In actual fact, if the elastic modulus of a weakly connected rubber is measured, the result is substantially larger than the value calculated from (3.19), indicating that entanglements do make a large contribution to the elastic modulus.

## 3.4 Swelling of gels

If a dried gel is placed in a solvent, the gel absorbs the solvent and its volume increases, sometimes up to several thousand times. This phenomenon is called gel swelling.

The driving force behind the swelling is the free energy of mixing of the polymer and solvent. If the polymer molecules making up the gel were not joined together, they would dissolve in the solvent to decrease the free energy of mixing, and eventually would be evenly distributed throughout the container. However, the polymers in a gel are connected, and so if the gel absorbs enough solvent it can attain an equilibrium state since there are two competing factors that determine its volume. One factor is the free energy of mixing of the gel and the solvent, which tries to increase the gel volume. The other factor is the change in elastic energy of the gel when the volume is varied, which acts to hinder the volume expansion.

The free energy of mixing of the gel and the solvent can be estimated from the Flory-Huggins theory. If we write the volume of the gel as $V$, we have

$$A_{\text{mix}} = \frac{V}{v_c} f_{\text{m}}(\phi) k_{\text{B}} T. \tag{3.67}$$

Here $f_{\text{m}}(\phi)$ is the free energy per lattice site, and is given by eqn (2.15). Keeping in mind that the molecular weight of the gel is very large $(N \sim 10^{20})$, we find

$$f_{\text{m}}(\phi) = (1 - \phi) \ln(1 - \phi) + \chi \phi(1 - \phi). \tag{3.68}$$

On the other hand, the elastic energy is given by (3.8). Writing $V_0$ for the volume of the gel before expansion and $\phi_0$ for the corresponding volume fraction of the polymer, the deformation gradient tensor can be written as

$$E_{\alpha\beta} = \left(\frac{V}{V_0}\right)^{1/3} \delta_{\alpha\beta} = \left(\frac{\phi_0}{\phi}\right)^{1/3} \delta_{\alpha\beta}. \tag{3.69}$$

Here we have used $V\phi = V_0\phi_0$. Therefore, the elastic energy becomes

$$A_{\text{el}} = \frac{3}{2} n_c k_{\text{B}} T \left(\frac{\phi_0}{\phi}\right)^{2/3}. \tag{3.70}$$

Thus, the free energy becomes

$$A = A_{\text{mix}} + A_{\text{el}} = \frac{V_0}{v_c} k_{\text{B}} T \left(\frac{\phi_0}{\phi} f_{\text{m}}(\phi) + \frac{3}{2} v_c \left(\frac{\phi_0}{\phi}\right)^{2/3}\right). \tag{3.71}$$

Here

$$\nu_c = \frac{n_c}{V_0} v_c. \tag{3.72}$$

The equilibrium volume of the gel, $V = V_0\phi_0/\phi$, is given by the $\phi$ which minimizes (3.71). That is, it is determined by the following equation:

$$\frac{\partial}{\partial \phi} \left[\frac{\phi_0}{\phi} f_{\text{m}}(\phi) + \frac{3}{2} \nu_c \left(\frac{\phi_0}{\phi}\right)^{2/3}\right] = 0. \tag{3.73}$$

Substituting (3.68) into (3.73) gives

$$\phi + \ln(1 - \phi) + \chi \phi^2 + \nu_c \left(\frac{\phi}{\phi_0}\right)^{1/3} = 0. \tag{3.74}$$

## 64 Polymer gels

The quantity $\chi$ depends only on the temperature and solvent , and so from (3.74) the volume fraction $\phi$ can be found.

According to (3.74), in the vicinity of $\chi = 0.5$, the volume fraction $\phi$ changes rapidly from 0 to 1, showing us the large change in the gel volume. This change is particularly dramatic for ionic gels, where the discontinuous volume change sometimes observed is called a volume phase transition. An example is shown in Fig. 3.8.

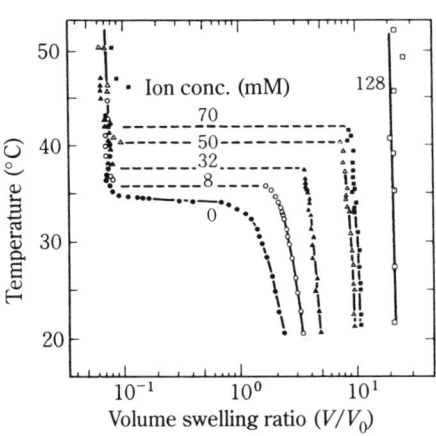

**Fig. 3.8** The variation of volume with temperature in an aqueous gel of cross-linked polyisopropylacrylamide. (Tanaka, T., and Hirokawa, N. (1986). *Kobunshi*, **35**, 237.)

# 4
# Molecular motion of polymers in dilute solution

A flexible polymer molecule in solution has an extremely large number of degrees of freedom, and its shape is incessantly changing. This perpetual molecular motion can be analysed through light scattering experiments on the solution, and recently it has become possible to observe directly the motion of very large molecules such as DNA using fluorescence microscopy (see Fig. 4.1). The understanding of molecular motion in polymer solutions enables us to explain many non-equilibrium phenomena, such as diffusion, dielectric relaxation, birefringence, etc.

In this chapter, we will investigate the molecular motion of a polymer in a dilute solution. However, our starting point will not be an equation of motion at the molecular level, since the polymer is such a complex many-bodied system and is also much larger than the solvent molecules. The usual approach is to divide the polymer into sections which are larger than the solvent molecules but smaller than the polymer overall, and consider their Brownian motion. The segments introduced in Chapter 1 are convenient

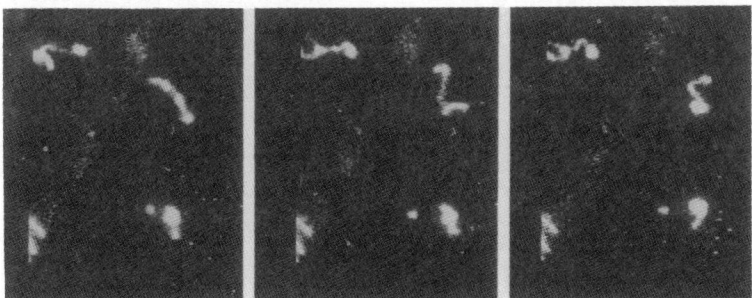

**Fig. 4.1** Photographs of DNA polymers in aqueous solution taken by fluorescence microscopy. There is a 1 s interval between successive frames. (Courtesy of Matsuzawa, Minagawa and Yoshikawa (Nagoya University).)

candidates for these sections. If we introduce a further simplification and assume that the segments are represented by small spheres called 'beads' which are linked together by springs, we have what is commonly known as the 'bead–spring model'. With this model, the problem of the molecular motion of the polymer is equivalent to the problem of a collection of interacting Brownian particles.

In this chapter, after a brief review of the general theory of Brownian motion, we will examine the motion of a polymer in dilute solution based on the bead–spring model.

## 4.1 General theory of Brownian motion

### 4.1.1 Brownian motion of spherical particles

Let us consider the Brownian motion of a spherical particle suspended in a solvent. We will consider its motion in the $x$ direction, and write $V(t)$ for the distribution function of the velocity in the $x$ direction at time $t$. This $V(t)$ is a randomly fluctuating function of time. The average of the product of the velocities at two different times, $\langle V(t_1)V(t_2)\rangle$, depends only on the time difference $t_1 - t_2$ if the system is in steady state, so we can write

$$\langle V(t_1)V(t_2)\rangle = C_{\mathrm{v}}(t_1 - t_2). \tag{4.1}$$

This is called the velocity correlation function.

At $t = 0$ the value of $C_{\mathrm{v}}(t)$ is the average squared velocity at equilibrium $\langle V^2 \rangle$, which equals $k_{\mathrm{B}}T/m$ for a particle of mass $m$. As $t$ increases, $C_{\mathrm{v}}(t)$ decreases and will eventually become zero. This time decay of $C_{\mathrm{v}}(t)$ is characterised by the velocity correlation time $\tau_{\mathrm{v}}$.

The velocity correlation time $\tau_{\mathrm{v}}$ can be estimated as follows. If the particle moves through the solvent with a velocity $V$, it experiences a viscous friction force $-\zeta V$ opposing its motion. Here $\zeta$ is the viscous friction constant, which becomes $\zeta = 6\pi\eta_{\mathrm{s}}a$ for a sphere of radius $a$ moving in a solvent of viscosity $\eta_{\mathrm{s}}$. Therefore, the equation of motion of the sphere can be written as follows:

$$m\frac{\mathrm{d}V}{\mathrm{d}t} = -\zeta V. \tag{4.2}$$

Equation (4.2) tells us that the velocity decays exponentially with the relaxation time $m/\zeta$, and so $\tau_{\mathrm{v}}$ can be estimated as

$$\tau_{\mathrm{v}} = m/\zeta. \tag{4.3}$$

Letting $\rho$ be the density of the sphere, we have $m = (4\pi/3)\rho a^3$, meaning that $\tau_{\mathrm{v}}$ is proportional to $\rho a^2/\eta_{\mathrm{s}}$. For $a = 10$ nm we find $\tau_{\mathrm{v}} \simeq 10^{-10}$ s, which is much shorter than the time scales usually considered in polymer dynamics($\sim 10^{-5}$ s).

Therefore from now on we will consider the limit of $\tau_v \to 0$. In this limit the velocity correlation function can be written as

$$\langle V(t)V(0)\rangle = 2D\delta(t). \tag{4.4}$$

The coefficient $D$ in (4.4) is actually the diffusion coefficient. To see this, let us consider the displacement of the particle during an interval of time $t$:

$$\xi(t) = \int_0^t dt' V(t'). \tag{4.5}$$

The average of the square of this, $\langle \xi(t)^2 \rangle$, can be calculated using (4.4) and (4.5) as follows:

$$\langle \xi(t)^2 \rangle = \int_0^t dt_1 \int_0^t dt_2 \langle V(t_1)V(t_2)\rangle$$
$$= \int_0^t dt_1 \int_0^t dt_2 2D\delta(t_1 - t_2)$$
$$= 2Dt. \tag{4.6}$$

Therefore, the average squared displacement is proportional to $t$, and so we see that $D$ does indeed correspond to the diffusion coefficient.

Even if the velocity correlation function cannot be written in the form of (4.4), the average squared displacement will always be proportional to $t$ for large $t$. The reason is that if the velocity correlation time is finite, the total displacement for times longer than $\tau_v$ is simply the sum of many smaller independent displacements of which there are $t/\tau_v$ in number. Therefore, in this case, the central limit theorem tells us that the displacement will have a Gaussian distribution. The variance is given by (4.6) , and the distribution $\psi(\xi, t)$ of the displacement $\xi$ is given by

$$\psi(\xi, t) = (4\pi Dt)^{-1/2} \exp\left(-\frac{\xi^2}{4Dt}\right). \tag{4.7}$$

### 4.1.2 The effect of a potential field

Next, let us put this particle undergoing Brownian motion into a potential field $U(x)$ which is a smoothly varying function of position. The particle experiences a force $-\partial U/\partial x$, and so begins to move through the solvent with the average velocity

$$\bar{V} = -\frac{1}{\zeta}\frac{\partial U}{\partial x}. \tag{4.8}$$

(Strictly speaking, the particle will reach the velocity given by (4.8) only after a time $\tau_v$ has passed, but here we are considering the limit $\tau_v \to 0$, and so we need not take into account this time delay.) If we assume that Brownian motion has negligible effect, (4.8) tells us that the particle displacement $x(t)$ satisfies the following equation:

$$\frac{dx}{dt} = -\frac{1}{\zeta}\frac{\partial U}{\partial x}. \tag{4.9}$$

According to (4.9), the particle moves in the direction of decreasing potential $U(x)$, and stops when it reaches the position of minimum potential.

However, if there is Brownian motion, the velocity of the particle fluctuates about the average value given by (4.8). To take into account these fluctuations, all we need to do is to add to the right hand side of (4.9) a probability function $g(t)$ which varies randomly with time :

$$\frac{dx}{dt} = -\frac{1}{\zeta}\frac{\partial U}{\partial x} + g(t). \tag{4.10}$$

Assuming that the velocity fluctuations in the presence of the potential field are the same as when there is no field, the mean and variance of $g(t)$ are as follows:

$$\langle g(t) \rangle = 0, \quad \langle g(t)g(t') \rangle = 2D\delta(t - t'). \tag{4.11}$$

Further, we know that we can assume $g(t)$ has a Gaussian distribution. Equation (4.10) is called the Langevin equation.[1]

### 4.1.3 Einstein's relation

Equations (4.10) and (4.11) are mathematical equations describing the Brownian motion of a particle in a potential field. Using this model, it is possible to simulate the Brownian motion on a computer. Integrating (4.10) from time $t$ to time $t + \Delta t$, and then discretizing with respect to time gives us

$$x(t + \Delta t) = x(t) - \frac{1}{\zeta}\frac{\partial U}{\partial x}\Delta t + \Delta G(t). \tag{4.13}$$

---

[1]The Langevin equation is often written with an inertial term included in (4.10):

$$m\frac{d^2 x}{dt^2} = -\zeta\frac{dx}{dt} - \frac{\partial U}{\partial x} + \zeta g(t). \tag{4.12}$$

However, in polymeric systems inertial terms can usually be dropped, and so we will use (4.10) as a starting point.

Here $\Delta G(t)$ is the probability distribution function formed by integrating $g(t)$ from $t$ to $t + \Delta t$:

$$\Delta G(t) = \int_t^{t+\Delta t} dt' g(t'). \tag{4.14}$$

The mean and variance of $\Delta G(t)$ are calculated from (4.11) as follows:

$$\langle \Delta G(t) \rangle = 0$$

$$\langle \Delta G(t) \Delta G(t') \rangle = \int_t^{t+\Delta t} dt_1 \int_{t'}^{t'+\Delta t} dt_2 2D\delta(t_1 - t_2)$$

$$= 2D\delta_{t,t'} \Delta t. \tag{4.15}$$

In order to carry out a simulation, all we need to do is to generate a Gaussian distribution random function $\Delta G(t)$ with a mean and variance given by (4.15), and use (4.13) to calculate the time evolution of $x(t)$.

If we start from an arbitrary configuration and carry out a simulation over a long time, we can calculate the probability that the position of a particle will lie between $x$ and $x + dx$. This probability distribution must be equal to the equilibrium Boltzmann distribution $\exp(-U/k_B T)$ for the above mathematical model to be physically valid. We will now see that this condition is satisfied if $D$ and $\zeta$ are related in a specific way.

As shown in Appendix 4.5.1, if $x(t)$ satisfies (4.13), the probability distribution function $\psi(x, t)$ of $x$ must satisfy the following equation:

$$\frac{\partial \psi}{\partial t} = \frac{\partial}{\partial x} \left[ \frac{1}{\zeta} \frac{\partial U}{\partial x} \psi + D \frac{\partial \psi}{\partial x} \right]. \tag{4.16}$$

The first term in (4.16) represents the effects of the average velocity, and the second term represents the diffusion due to Brownian motion. As $t \to \infty$, the solution of (4.16) approaches the following steady state:

$$\psi(x) \propto \exp\left( -\frac{U}{D\zeta} \right). \tag{4.17}$$

Equation (4.17) coincides with the Boltzmann distribution if the following relation holds:

$$D = \frac{k_B T}{\zeta}. \tag{4.18}$$

Equation (4.18) is called the Einstein relation.

The Einstein relation tells us that a physical quantity which describes the thermal fluctuations in a system (the diffusion coefficient) can be calculated from a quantity that describes the response of the system under an external force (the viscosity coefficient). There is a generalisation of this relation called the fluctuation–dissipation theorem.

Using Einstein's relation, the distribution function can be seen to satisfy the following equation:

$$\frac{\partial \psi}{\partial t} = \frac{\partial}{\partial x} D \left[ \frac{\partial \psi}{\partial x} + \frac{1}{k_B T} \frac{\partial U}{\partial x} \psi \right]. \tag{4.19}$$

The above arguments can be generalised for systems with many degrees of freedom. Ignoring fluctuations, let us assume that the variables $\{x_i\} = (x_1, x_2, \cdots x_n)$ satisfy the following time-evolution equation:

$$\frac{dx_i}{dt} = -\sum_j \mu_{ij} \frac{\partial U}{\partial x_j}. \tag{4.20}$$

Here $U(\{x_i\})$ is the potential energy which determines the equilibrium state of the system. The coefficients $\mu_{ij}$, called the mobility matrix, are generally functions of $\{x_i\}$. Our next step is to incorporate Brownian motion into (4.20). The time evolution equation for the distribution function becomes the following, with the equilibrium distribution function equaling $\exp(-U/k_B T)$:

$$\frac{\partial \psi}{\partial t} = \sum_{ij} \frac{\partial}{\partial x_i} D_{ij} \left[ \frac{\partial \psi}{\partial x_j} + \frac{1}{k_B T} \frac{\partial U}{\partial x_j} \psi \right]. \tag{4.21}$$

Here

$$D_{ij} = \mu_{ij} k_B T. \tag{4.22}$$

The Langevin equation corresponding to (4.21) is

$$\frac{dx_i}{dt} = -\sum_j \mu_{ij} \frac{\partial U}{\partial x_j} + g_i(t) + \sum_j \frac{\partial D_{ij}}{\partial x_j}. \tag{4.23}$$

Here the mean and variance of $g_i$ are as follows:

$$\langle g_i(t) \rangle = 0 \qquad \langle g_i(t) g_j(t') \rangle = 2D_{ij}\delta(t - t'). \tag{4.24}$$

*4.1.4 The Brownian motion of a harmonic oscillator*

As a simple example, let us consider the Brownian motion of a harmonic oscillator with the following potential:

$$U(x) = \frac{1}{2}kx^2. \tag{4.25}$$

The following equation governs the behaviour of $x(t)$:

$$\frac{dx}{dt} = -\frac{k}{\zeta}x + g(t). \tag{4.26}$$

Solving (4.26) for $x(t)$ gives us

$$x(t) = \int_{-\infty}^{t} dt' e^{-k(t-t')/\zeta} g(t').$$  (4.27)

Therefore, the time correlation function of $x(t)$ can be calculated as follows:

$$\langle x(t)x(0)\rangle = \int_{-\infty}^{t} dt_1 \int_{-\infty}^{0} dt_2 \exp[-k(t-t_1-t_2)/\zeta]\langle g(t_1)g(t_2)\rangle.$$  (4.28)

Using (4.11) and (4.18) gives us

$$\langle x(t)x(0)\rangle = \int_{-\infty}^{t} dt_1 \int_{-\infty}^{0} dt_2 \exp[-k(t-t_1-t_2)/\zeta]\frac{2k_B T}{\zeta}\delta(t_1-t_2)$$

$$= \frac{k_B T}{k}\exp(-t/\tau).$$  (4.29)

Here

$$\tau = \frac{\zeta}{k}.$$  (4.30)

If we let $t = 0$ in (4.29) we get $\langle x^2\rangle = k_B T/k$. This is the same as the value calculated using the Boltzmann distribution $\psi_{eq} \propto \exp(-kx^2/2k_B T)$. It is the Einstein relation which guarantees this agreement.

The average squared displacement $\langle (x(t) - x(0))^2\rangle$ after a time $t$ can be easily calculated from $\langle x(t)x(0)\rangle$.

$$\langle (x(t) - x(0))^2\rangle = \langle x(t)^2\rangle + \langle x(0)^2\rangle - 2\langle x(t)x(0)\rangle$$

$$= 2\langle x^2\rangle - 2\langle x(t)x(0)\rangle$$

$$= \frac{2k_B T}{k}(1 - \exp(-t/\tau)).$$  (4.31)

In particular, if we consider the limit $t \to 0$, eqn (4.31) gives us

$$\langle (x(t) - x(0))^2\rangle = \frac{2k_B T}{\zeta}t = 2Dt.$$  (4.32)

As expected, this agrees with (4.6).

## 4.2 The bead–spring model

### 4.2.1 Rouse theory

Now let us consider the Brownian motion of a polymer molecule. We will use the bead–spring model shown in Fig. 1.3. If we assume that the beads experience a drag force proportional to their velocity as they move through the solvent, then the position of the beads $R_n(t)$ will satisfy the following Langevin equation:

$$\frac{dR_n}{dt} = -\frac{1}{\zeta}\frac{\partial U}{\partial R_n} + g_n. \tag{4.33}$$

Here $\zeta$ is the friction coefficient of a bead.

Using the potential (1.19), the governing equations of this model can be written as follows. For $n = 1, 2, ..., (N-1)$,

$$\frac{dR_n}{dt} = \frac{k}{\zeta}(R_{n+1} + R_{n-1} - 2R_n) + g_n. \tag{4.34}$$

For $n = 0$ and $n = N$,

$$\frac{dR_0}{dt} = \frac{k}{\zeta}(R_1 - R_0) + g_0 \tag{4.35}$$

$$\frac{dR_N}{dt} = \frac{k}{\zeta}(R_{N-1} - R_N) + g_N. \tag{4.36}$$

If we define $R_{-1}$, and $R_{N+1}$ as

$$R_{-1} \equiv R_0, \qquad R_{N+1} \equiv R_N, \tag{4.37}$$

then (4.35) and (4.36) can be included in (4.34). This model was first proposed by P.E. Rouse, and so is called the Rouse model.

In order to proceed with our calculations, it is convenient to assume that the beads are continuously distributed along the polymer chain. Letting $n$ be a continuous variable, and writing $R_n(t)$ as $R(n, t)$, eqn (4.34) takes the following form:

$$\frac{\partial R}{\partial t} = \frac{k}{\zeta}\frac{\partial^2 R}{\partial n^2} + g(n, t). \tag{4.38}$$

Further, the conditions of (4.37) become the following boundary conditions at $n = 0$ and $n = N$:

$$\frac{\partial R}{\partial n} = 0. \tag{4.39}$$

Equation (4.38) has the form of a linear harmonic oscillator, and so if we introduce normalized coordinates we can decompose the motion into independent modes. Considering (4.39), we introduce the following normalized coordinates:

$$X_p(t) = \frac{1}{N}\int_0^N dn \cos\left(\frac{p\pi n}{N}\right)R(n, t) \quad p = 0, 1, 2, ... \tag{4.40}$$

Using this, (4.38) can be rewritten as follows:

$$\frac{dX_p}{dt} = -\frac{k_p}{\zeta_p}X_p + g_p. \tag{4.41}$$

Here

$$\zeta_0 = N\zeta \ , \ \zeta_p = 2N\zeta \ , \ k_p = \frac{2p^2\pi^2 k}{N} = \frac{6\pi^2 k_B T}{Nb^2} p^2. \tag{4.42}$$

The random force $g_p(t)$ has a mean of 0, and a variance given by

$$\langle g_{p\alpha}(t)g_{q\beta}(t')\rangle = 2\delta_{pq}\delta_{\alpha\beta}\frac{k_B T}{\zeta_p}\delta(t - t'). \tag{4.43}$$

Using the results of Section 4.1.4, the correlation function of the normalized coordinates can be calculated as follows:

$$\langle (X_0(t) - X_0(0))_\alpha (X_0(t) - X_0(0))_\beta \rangle = \delta_{\alpha\beta}\frac{2k_B T}{\zeta_0}t. \tag{4.44}$$

$$\langle X_{p\alpha}(t)X_{q\beta}(0)\rangle = \delta_{pq}\delta_{\alpha\beta}\frac{k_B T}{k_p}e^{-t/\tau_p} \quad (p = 1, 2, \ldots) \tag{4.45}$$

Here

$$\tau_p = \frac{\zeta_p}{k_p} = \frac{\tau_1}{p^2} = \frac{1}{p^2}\frac{\zeta N^2 b^2}{3\pi^2 k_B T}. \tag{4.46}$$

*4.2.2 Features of Brownian motion of the polymer*

Let us investigate some of the features of the motion of the Rouse model, using eqns (4.44)–(4.46).

*Motion of the centre of mass*
The position of the centre of mass

$$R_G(t) = \frac{1}{N}\int_0^N dn R(n, t). \tag{4.47}$$

is the same as the normal coordinate $X_0(t)$. Therefore, the average squared displacement of the centre of mass can be found from (4.44) as follows:

$$\langle (R_G(t) - R_G(0))^2 \rangle = 3\frac{2k_B T}{\zeta_0}t = \frac{6k_B T}{N\zeta}t. \tag{4.48}$$

Thus, the centre of mass undergoes diffusion with a diffusion constant

$$D_G = \frac{k_B T}{N\zeta}. \tag{4.49}$$

Note that $D_G$ is inversely proportional to $N$.

*Rotational motion*

To characterize the rotational motion of the polymer molecule as a whole, let us consider the time correlation function $\langle P(t) \cdot P(0) \rangle$ of the end-to-end vector $P$. Using normalized coordinates, $P(t)$ can be written as

$$P(t) = R(N, t) - R(0, t) = -4 \sum_p X_p(t), \tag{4.50}$$

where $p$ ranges over positive odd integers. Therefore, using (4.45) we have

$$\langle P(t) \cdot P(0) \rangle = 16 \sum_p \frac{3k_B T}{k_p} \exp(-t/\tau_p)$$

$$= Nb^2 \sum_p \frac{8}{p^2 \pi^2} \exp(-tp^2/\tau_1), \tag{4.51}$$

where $p$ similarly ranges over positive odd integers.

We see that $\langle P(t) \cdot P(0) \rangle$ is a summation of many terms with different relaxation times. However, the coefficient of most of the terms decreases rapidly with $p$ and so, to a very close approximation, $\langle P(t) \cdot P(0) \rangle$ decays exponentially with a single relaxation time $\tau_1$. This relaxation time $\tau_1$ is called the rotational relaxation time, and is written as $\tau_r$. From (4.46) we have

$$\tau_r = \frac{\zeta N^2 b^2}{3\pi^2 k_B T}. \tag{4.52}$$

Since $\tau_r \simeq Nb^2/D_G$, we see that the rotational relaxation time is also equal to the time required for the centre of mass of the polymer to diffuse a distance comparable to the size of the polymer.

*Motion of the segments*

To study the internal motion of the polymer chain, let us consider the average squared displacement of the $n$th segment over a time $t$:

$$\phi(n, t) \equiv \langle (R(n, t) - R(n, 0))^2 \rangle. \tag{4.53}$$

From (4.40), $R(n, t)$ can be written as follows using normalized coordinates:

$$R(n, t) = X_0(t) + 2 \sum_{p=1}^{\infty} \cos\left(\frac{p\pi n}{N}\right) X_p(t). \tag{4.54}$$

Therefore, using (4.44) and (4.45), $\phi(n, t)$ can be calculated as follows:

$$\phi(n, t) = 6D_G t + \frac{4Nb^2}{\pi^2} \sum_{p=1}^{\infty} \cos^2\left(\frac{p\pi n}{N}\right) \frac{1}{p^2} \left(1 - \exp(-tp^2/\tau_r)\right). \tag{4.55}$$

For $t \gg \tau_r$, $\phi(n, t) \simeq 6D_G t$ and the displacement of the segments is determined by the diffusion constant of the centre of mass. On the other hand, for $t \ll \tau_r$, the motion of the segments reflects the internal motion due to the many modes of vibration. For $t \ll \tau_r$, the summation in (4.55) can be rewritten as an integral, and if we replace $\cos^2(p\pi n/N)$ by its average value $1/2$ we have

$$
\phi(n, t) = 6D_G t + \frac{4Nb^2}{\pi^2} \int_0^\infty dp \frac{1}{2p^2} \left(1 - \exp(-tp^2/\tau_r)\right)
$$

$$
= \frac{2Nb^2}{\pi^2} \left[\frac{t}{\tau_r} + \left(\frac{\pi t}{\tau_r}\right)^{1/2}\right] \approx \frac{2}{\pi^{3/2}} Nb^2 \left(\frac{t}{\tau_r}\right)^{1/2} \tag{4.56}
$$

Thus we see that for $t \ll \tau_r$, the average squared displacement increases proportionally to $t^{1/2}$.

### 4.2.3 Comparison with experiments

The Rouse model may seem to be a very natural way to describe the Brownian motion of a polymer chain, but unfortunately its conclusions do not agree with experiments. As can be seen from (4.49) and (4.52), in the Rouse model the diffusion constant of the centre of mass and the rotational relaxation time depend on the polymer molecular weight $M$ as follows:

$$
D_G \propto M^{-1}, \quad \tau_r \propto M^2. \tag{4.57}
$$

However, the following dependencies have been measured experimentally:

$$
D_G \propto M^{-\nu}, \quad \tau_r \propto M^{3\nu}. \tag{4.58}
$$

Here, the exponent $\nu$ is that which is used to express the dependence of the polymer size $R_g$ on molecular weight. (In the $\Theta$ state $\nu = 1/2$, and in a good solvent $\nu \simeq 3/5$.) The reason for this discrepency between experiment and the Rouse model is that in the latter we have assumed the average velocity of a particular bead is determined only by the external force acting on it, and is independent of the motion of the other beads. However, in reality the motion of one bead is influenced by the motion of the surrounding beads through the medium of the solvent. For example, if one bead moves the solvent surrounding it will also move, and as a result other beads will be dragged along. This type of interaction transmitted by the motion of the solvent is called hydrodynamic interaction. We will discuss this in the next section.

## 4.3 Hydrodynamic interactions

### 4.3.1 The mobility matrix for a many-particle system

Hydrodynamic interactions do not change the potential $U$ in (4.20), but they do alter the mobility matrix $\mu_{ij}$.

To determine the mobility matrix in a viscous solvent, we shall proceed as follows. As shown in Fig. 4.2, we assume that several spheres of radius $a$ are suspended in a quiescent viscous fluid. Let us write $F_n (n = 1, 2, ...)$ for the external force acting on each sphere, and assume that each sphere moves with the steady velocity $V_n$. If the forces are weak, we can assume that there is a linear relationship between them and $V_n$, and so the following relation holds:

$$V_n = \sum_m H_{nm} \cdot F_m,\tag{4.59}$$

where the tensor $H_{nm}$ corresponds to the mobility matrix.

In order to calculate $H_{nm}$, we will follow a similar procedure to that used when we derived the Stokes law, and solve the hydrodynamic equations. Since the flow is slow, we can drop the inertial terms, and so the fluid velocity $v(r)$ must satisfy the following equations:

$$\eta_s \nabla^2 v + \nabla p = 0.\tag{4.60}$$
$$\nabla \cdot v = 0.\tag{4.61}$$

Assuming that the velocities of all the spheres $V_n$ are given, we can solve the above equations under the boundary condition that the velocity of the fluid at the surface of each sphere equals the velocity of the sphere. From this, we can calculate the viscous resistance force acting on each sphere. From the

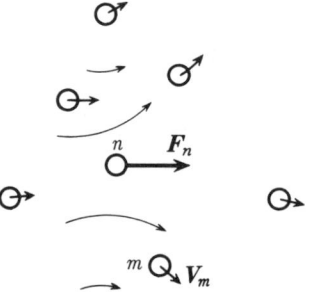

**Fig. 4.2** The hydrodynamic interaction. If bead $n$ moves under the action of the force $F_n$, a flow is created in the surrounding fluid, which causes the other beads to move.

condition that this viscous resistance force equals the external force applied to the sphere, $F_n$, the relationship between the sphere velocity and the external force can be found. In practice, this is a very complex calculation to carry out, but since (4.60) and (4.61) are linear equations, the solution will definitely take the linear form of (4.59).

Since $H_{nm}$ depends on the positions of all the spheres, calculating it is in general extremely difficult. However, if the average distance between neighbouring spheres is much larger than the sphere radius, the following approximation can be made.

Let us focus attention on the $n$th sphere. Assuming that this sphere is in a flow field $v'(r)$ which varies gradually with position, we can assume that the viscous resistance force is given by $-6\pi\eta_s a(V_n - v'(R_n))$ . Thus, from the balance of forces condition, we have

$$F_n = 6\pi\eta_s a(V_n - v'(R_n)).\qquad(4.62)$$

Now, $v'(r)$ can be thought of as the flow field created by the motion of spheres other than the $n$th one. In this case, if we ignore the finite size of the other spheres and assume that a force $F_m$ is applied at their centres, $v'(r)$ will satisfy the following equations:

$$\eta_s \nabla^2 v' + \nabla p = \sum_m{}' \delta(r - R_m)\, F_m, \quad \nabla \cdot v' = 0.\qquad(4.63)$$

Here $\sum'$ means summation over spheres other than the $n$th one. Equation (4.63) can be solved using Fourier transforms, with the following result:

$$v'(r) = \sum_m{}' T(r - R_m) \cdot F_m.\qquad(4.64)$$

Here

$$T(r) = \frac{1}{8\pi\eta_s r}\left[\frac{rr}{r^2} + I\right].\qquad(4.65)$$

is called the Oseen tensor. From (4.62) and (4.64) we have

$$V_n = \frac{F_n}{6\pi\eta_s a} + \sum_m{}' T(R_n - R_m) \cdot F_m.\qquad(4.66)$$

Therefore, the end result of this calculation gives the tensor $H_{nm}$ as

$$H_{nm} - \begin{cases} I/(6\pi\eta_s a) & n = m \\ T(R_n - R_m) & n \neq m. \end{cases}\qquad(4.67)$$

If we take into account hydrodynamic interactions, the basic equation describing the Brownian motion becomes

$$\frac{d\mathbf{R}_n}{dt} = -\sum_m \mathbf{H}_{nm} \cdot \frac{\partial U}{\partial \mathbf{R}_m} + \mathbf{g}_n + \sum_m \frac{\partial}{\partial \mathbf{R}_m} \cdot \mathbf{H}_{nm} k_B T. \tag{4.68}$$

Using (4.65), the final term becomes 0, and so we can write the basic equation as

$$\frac{d\mathbf{R}_n}{dt} = -\sum_m \mathbf{H}_{nm} \cdot \frac{\partial U}{\partial \mathbf{R}_m} + \mathbf{g}_n. \tag{4.69}$$

The mean and variance of the random force $\mathbf{g}_n(t)$ in (4.69) are given by

$$\langle g_{n\alpha}(t) \rangle = 0, \quad \langle g_{n\alpha}(t) g_{m\beta}(t') \rangle = 2(\mathbf{H}_{nm})_{\alpha\beta} k_B T \delta(t - t'). \tag{4.70}$$

### 4.3.2 Zimm theory

Zimm modified the bead–spring model of a polymer to include hydrodynamic interaction effects. In this case, the equation of motion of the bead–spring polymer (4.69) becomes the following:

$$\frac{d\mathbf{R}_n}{dt} = k \sum_m \mathbf{H}_{nm} \cdot (\mathbf{R}_{m+1} + \mathbf{R}_{m-1} - 2\mathbf{R}_m) + \mathbf{g}_n. \tag{4.71}$$

Since $\mathbf{H}_{nm}$ depends on $\mathbf{R}_n$, eqn (4.71) is a non-linear equation in $\mathbf{R}_n(t)$, and is almost impossible to solve. Zimm's idea was to replace $\mathbf{H}_{nm}$ (the factor causing the non-linearity) by its equilibrium average value $\langle \mathbf{H}_{nm} \rangle_{eq}$. This is called the preaveraging approximation. At first this may appear to be a very rough approach, but as we shall see later, the preaveraging approximation turns out to be quite good because of the long range nature of the hydrodynamic interactions. For example, if we calculate the diffusion constant using this approximation, the result is within 10% of the value calculated rigorously.

To find $\langle \mathbf{H}_{nm} \rangle_{eq}$, let us consider the $\Theta$ state. Since $\mathbf{H}_{nm}$ is a function only of $\mathbf{r}_{nm} \equiv \mathbf{R}_n - \mathbf{R}_m$, and $\mathbf{r}_{nm}$ has a Gaussian distribution with a variance of $|n - m| b^2$, we can calculate $\langle \mathbf{H}_{nm} \rangle_{eq}$ as follows:

$$\langle \mathbf{H}_{nm} \rangle_{eq} = \int d\mathbf{r} \left( \frac{3}{2\pi |n - m| b^2} \right)^{3/2} \exp\left( -\frac{3r^2}{2|n - m| b^2} \right) \frac{1}{8\pi \eta_s r} \left( \frac{\mathbf{rr}}{r^2} + \mathbf{I} \right). \tag{4.72}$$

Taking an average over $\mathbf{r}$, and noting that the average of $\mathbf{rr}/r^2$ is $\mathbf{I}/3$ (where $\mathbf{I}$ is the unit tensor), we have

$$\langle \mathbf{H}_{nm} \rangle_{eq} = \int_0^\infty dr\, 4\pi r^2 \left( \frac{3}{2\pi |n - m| b^2} \right)^{3/2} \exp\left( -\frac{3r^2}{2|n - m| b^2} \right) \frac{1}{8\pi \eta_s r} \frac{4}{3} \mathbf{I}$$
$$\equiv h(n - m)\mathbf{I}. \tag{4.73}$$

Here

$$h(n - m) = \frac{1}{(6\pi^3 \mid n - m \mid)^{1/2}\eta_s b}.$$ (4.74)

Therefore the equation of motion of $R(n, t)$ with this continuous model becomes

$$\frac{\partial R(n, t)}{\partial t} = k \int_0^N dm \, h(n - m) \frac{\partial^2 R(m, t)}{\partial m^2} + g(n, t).$$ (4.75)

Here the correlation function of $g(n, t)$ is given by

$$\langle g_\alpha(n, t)g_\beta(m, t')\rangle = 2h(n - m)k_B T \delta_{\alpha\beta}\delta(t - t').$$ (4.76)

Rewriting (4.75) using the normalized coordinates of the Rouse model we have

$$\frac{dX_p}{dt} = -\sum_q h_{pq} k_q X_q + g_p.$$ (4.77)

Here $k_q$ is given by (4.42), and $h_{pq}$ is defined as follows:

$$h_{pq} = \frac{1}{N^2} \int_0^N dn \int_0^N dm \cos\left(\frac{p\pi n}{N}\right) \cos\left(\frac{q\pi m}{N}\right) h(n - m).$$ (4.78)

Setting $n - m = l$, (4.78) can be rewritten as

$$\begin{aligned}
h_{pq} &= \frac{1}{N^2} \int_0^N dn \int_{-n}^{N-n} dl \cos\left(\frac{p\pi n}{N}\right) \cos\left(\frac{q\pi(l + n)}{N}\right) h(l) \\
&= \frac{1}{N^2} \int_0^N dn \left[ \cos\left(\frac{p\pi n}{N}\right) \cos\left(\frac{q\pi n}{N}\right) \underline{\int_{-n}^{N-n} dl h(l) \cos\left(\frac{q\pi l}{N}\right)} \right. \\
&\quad \left. - \cos\left(\frac{p\pi n}{N}\right) \sin\left(\frac{q\pi n}{N}\right) \underline{\int_{-n}^{N-n} dl h(l) \sin\left(\frac{q\pi l}{N}\right)} \right].
\end{aligned}$$ (4.79)

For large $q$, the underlined terms rapidly approach the following integrals:

$$\int_{-\infty}^{\infty} dl \, h(l) \cos\left(\frac{q\pi l}{N}\right) = \frac{\sqrt{N}}{(3\pi^3 q)^{1/2}\eta_s b}$$

$$\int_{-\infty}^{\infty} dl \, h(l) \sin\left(\frac{q\pi l}{N}\right) = 0.$$ (4.80)

With this substitution $h_{pq}$ becomes

$$h_{pq} = \frac{\sqrt{N}}{(3\pi^3 p)^{1/2}\eta_s b} \frac{1}{2N} \delta_{pq}.$$ (4.81)

Therefore (4.77) can be written in the same form as the Rouse model:

$$\frac{dX_p}{dt} = -\frac{k_p}{\zeta_p} X_p + g_p.$$ (4.82)

Here $\zeta_p \equiv 1/h_{pp}$, and

$$\zeta_p = (12\pi^3)^{1/2}(Nb^2p)^{1/2}\eta_s.$$ (4.83)

$$k_p = \frac{6\pi^2 k_B T}{Nb^2}p^2.$$ (4.84)

If $p$ is small we cannot make a substitution like (4.81). In particular, if $p = 0$, eqn (4.83) is not correct. However, $\zeta_0$ can be calculated from $h_{00}$ as follows:

$$\zeta_0 = \left[\frac{1}{N^2}\int_0^N dn \int_0^N dm\, h(n-m)\right]^{-1}$$ (4.85)

$$= \frac{3}{8}(6\pi^3)^{1/2}\eta_s b\sqrt{N}.$$ (4.86)

For $p = 1$, the difference between (4.83) and $h_{11}^{-1}$ is not large, and so we will use (4.83) and (4.86) from now on.

Since the equation for the normal modes is the same as that for the Rouse model, we can immediately write the equations for the diffusion constant of the centre of mass and the rotational relaxation time using the results of the previous section:

$$D_G = \frac{k_B T}{\zeta_0} = \frac{8k_B T}{3(6\pi^3)^{1/2}\eta_s\sqrt{Nb}} = 0.196\frac{k_B T}{\eta_s\sqrt{Nb}}$$ (4.87)

$$\tau_r = \frac{\zeta_1}{k_1} = \frac{\eta_s(\sqrt{Nb})^3}{\sqrt{3\pi}k_B T} = 0.325\frac{\eta_s(\sqrt{Nb})^3}{k_B T}.$$ (4.88)

Therefore $D_G$ and $\tau_r$ depend on the molecular weight $M$ as follows:

$$D_G \propto M^{-1/2}, \quad \tau_r \propto M^{3/2}.$$ (4.89)

The dependence of these quantities on the molecular weight agrees with experiments performed on solutions in the $\Theta$ state.

Further, the relaxation times of the normal modes are

$$\tau_p = \frac{\zeta_p}{k_p} = \tau_r p^{-3/2}.$$ (4.90)

Therefore, the average squared displacement of the segments is

$$\phi(n,t) = \frac{2Nb^2}{\pi^2}\int_0^\infty dp\,\frac{1}{p^2}\left(1 - \exp(-tp^{3/2}/\tau_r)\right)$$

$$= \frac{2}{\pi^2}\Gamma(\tfrac{1}{3})Nb^2\left(\frac{t}{\tau_r}\right)^{2/3}.$$ (4.91)

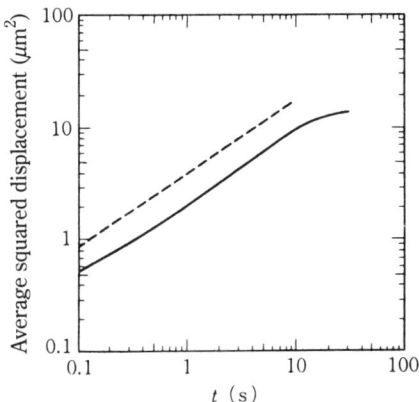

**Fig. 4.3** The average squared displacement of the terminal segment of a DNA molecule, observed by fluorescence microscopy. The dashed line is calculated from the theory of Zimm. (Matsumoto, M., Sakaguti, T., Kimura, H., Doi, M., Matsuzawa, Y., Minagawa, K., and Yoshikawa, K. (1991). *J. Polym. Sci.*, **30**, 779, Fig. 5.)

The relationship $\phi(n, t) \propto t^{2/3}$ has been confirmed by analysis of the Brownian motion of DNA molecules (Fig. 4.3).

The above discussion has been confined to solutions in the $\Theta$ state. For details on the calculation of the diffusion constant in good solvents, the reader should consult the references listed at the end of this book.

### 4.3.3 The dynamic scaling law

The results of the Zimm theory show that quantities which describe the motion of the polymer such as $D_G$ and $\tau_r$ also satisfy the principle of scaling. Actually, if we look at (4.87) and (4.88) , we see that $D_G$ and $\tau_r$ are determined only by the polymer radius of gyration $R_g = (Nb^2/6)^{1/2}$ which is a macroscopic quantity, and that quantities like $N$ and $b$ do not enter explicitly into the equations. In other words, the scaling principle explained in Section 1.4 also holds true for dynamic quantities.

As an example of this dynamic scaling law, let us consider the average squared displacement of the terminal segment $\phi(t) \equiv \langle (\boldsymbol{R}(0, t) - \boldsymbol{R}(0, 0))^2 \rangle$. The function $\phi(t)$ should depend only on quantities such as $N, b, k_B T, \eta_s$, and so from dimensional analysis we have

$$\phi(t) = b^2 f\left(N, \frac{k_B T t}{\eta_s b^3}\right). \tag{4.92}$$

Since this is invariant under the transformation $N \to \lambda^{-1} N, b \to \lambda^\nu b$, we must have

$$\phi(t) = (N^\nu b)^2 f\left(\frac{k_\mathrm{B} T t}{\eta_\mathrm{s}(N^\nu b)^3}\right). \tag{4.93}$$

This can be rewritten as follows:

$$\phi(t) = (R_\mathrm{g})^2 f\left(\frac{t}{\tau_\mathrm{r}}\right). \tag{4.94}$$

Here

$$\tau_\mathrm{r} \simeq \frac{\eta_\mathrm{s} R_\mathrm{g}^3}{k_\mathrm{B} T}. \tag{4.95}$$

is the rotational relaxation time of the polymer.

From (4.94) we can draw several important conclusions. For $t \gg \tau_\mathrm{r}$, we know that $\phi(t)$ should be proportional to $t$, which gives us

$$\phi(t) \simeq R_\mathrm{g}^2 \frac{t}{\tau_\mathrm{r}} \simeq \frac{k_\mathrm{B} T}{\eta_\mathrm{s} R_\mathrm{g}} t. \tag{4.96}$$

From this, we see that the diffusion constant is given by

$$D_\mathrm{G} \simeq \frac{k_\mathrm{B} T}{\eta_\mathrm{s} R_\mathrm{g}}. \tag{4.97}$$

On the other hand, if $t \ll \tau_\mathrm{r}$, then $\phi(t)$ should be independent of molecular weight. If we let $R_\mathrm{g} \propto N^\nu$ and $\tau_\mathrm{r} \propto N^{3\nu}$ in (4.94), then for $\phi(t)$ to be independent of $N$ we must have

$$\phi(t) \simeq R_\mathrm{g}^2 \left(\frac{t}{\tau_\mathrm{r}}\right)^{2/3} \propto t^{2/3}. \tag{4.98}$$

This agrees with our previous result (4.91). The result $\phi(t) \propto t^{2/3}$ is independent of $\nu$, and so holds for both $\Theta$ solvents and good solvents.

## 4.4. Dynamic light scattering

### 4.4.1 Dynamic structure factor

If the target object moves around during a light scattering experiment, the intensity of the scattered light will fluctuate with time. From the time correlation of this fluctuation, a dynamic structure factor can be defined as follows:

$$g(q, t) = \frac{1}{N} \sum_{n,m} \left\langle \exp[iq \cdot (R_n(t) - R_m(0))] \right\rangle. \qquad (4.99)$$

The diffusion constant of a polymer in a dilute solution can be found from the dynamic structure factor. Let us consider the region of small scattering angle $qR_g \ll 1$. Since $| R_n(t) - R_G(t) | \leq R_g$, for $qR_g \ll 1$ we can replace $R_n(t)$ and $R_m(0)$ in (4.99) by $R_G(t)$ and $R_G(0)$, respectively. Then (4.99) becomes

$$g(q, t) = N \left\langle \exp[iq \cdot (R_G(t) - R_G(0))] \right\rangle. \qquad (4.100)$$

For large $t$, the vector $R_G(t) - R_G(0)$ has a Gaussian distribution with variance $2D_G t$, and so the average in (4.100) becomes

$$g(q, t) = N \int dr \exp[iq \cdot r](4\pi D_G t)^{-3/2} \exp\left(-\frac{r^2}{4D_G t}\right)$$
$$= N \exp(-D_G q^2 t). \qquad (4.101)$$

Therefore, we see that $g(q, t)$ decreases exponentially, and from the rate of decay we can calculate the diffusion constant $D_G$ of the polymer.

*4.4.2 Initial decay rate of $g(q, t)$*

For arbitrary $q$ it is difficult to calculate rigorously the dynamic structure factor, but the initial rate of decay

$$\Gamma_q = -\left[\frac{d \ln g(q, t)}{dt}\right]_{t=0}. \qquad (4.102)$$

can be calculated relatively easily. Substituting

$$c_q(\{R_n\}) = \sum_n \exp(iq \cdot R_n), \qquad (4.103)$$

we see that $g(q, t)$ can be written as

$$g(q, t) = \frac{1}{N} \left\langle c_q(\{R_n(t)\}) c_{-q}(\{R_n(0)\}) \right\rangle. \qquad (4.104)$$

As shown in Appendix 4.5.2, when $R_n(t)$ satisfies (4.69) the rate of decay of $g(q, t)$ can be calculated as follows:

$$\frac{d}{dt} g(q, t)|_{t=0} = -\frac{1}{N} \sum_{n,m} k_B T \left\langle \frac{\partial c_q}{\partial R_n} \cdot H_{nm} \cdot \frac{\partial c_{-q}}{\partial R_m} \right\rangle \qquad (4.105)$$

$$= -\frac{1}{N} \sum_{n,m} k_B T \langle qq : H_{nm} \exp\{iq \cdot (R_n - R_m)\}\rangle. \qquad (4.106)$$

Using the Fourier transform of the Oseen tensor, $H_{nm}$ can be written as follows:

$$H_{nm} = \frac{1}{\eta_s} \int \frac{d\mathbf{k}}{(2\pi)^3} \frac{\mathbf{I} - \hat{\mathbf{k}}\hat{\mathbf{k}}}{k^2} \exp[i\mathbf{k} \cdot (\mathbf{R}_n - \mathbf{R}_m)]. \tag{4.107}$$

Thus

$$\frac{d}{dt} g(\mathbf{q}, t)|_{t=0} = -\frac{k_B T}{\eta_s} \int \frac{d\mathbf{k}}{(2\pi)^3} \frac{q^2 - (\hat{\mathbf{k}} \cdot \mathbf{q})^2}{k^2} g(\mathbf{q} + \mathbf{k}). \tag{4.108}$$

Here

$$g(\mathbf{q}) = \frac{1}{N} \sum_{n,m} \langle \exp[i\mathbf{q} \cdot (\mathbf{R}_n - \mathbf{R}_m)] \rangle \tag{4.109}$$

is the pair correlation function introduced in Chapter 1.

Using (4.108), eqn (4.102) becomes

$$\Gamma_q = \frac{k_B T}{\eta_s} \int \frac{d\mathbf{k}}{(2\pi)^3} \frac{q^2 - (\hat{\mathbf{k}} \cdot \mathbf{q})^2}{k^2} \frac{g(\mathbf{k} + \mathbf{q})}{g(\mathbf{q})}. \tag{4.110}$$

If $g(\mathbf{q})$ is given by eqn (1.38), we obtain the following result after a rather laborious calculation:

$$\Gamma_q = \frac{k_B T}{6\pi\eta_s\xi} q^2 F(q\xi). \tag{4.111}$$

Here

$$F(x) = \frac{3}{4} \frac{1 + x^2}{x^3} [x + (x^2 - 1)\tan^{-1} x]. \tag{4.112}$$

The asymptotes of this function are given by

$$\Gamma_q = \begin{cases} \dfrac{k_B T}{6\pi\eta_s\xi} q^2 & (q\xi \ll 1) \\[2mm] \dfrac{k_B T}{16\eta_s} q^3 & (q\xi \gg 1). \end{cases} \tag{4.113}$$

For $qR_g \gg 1$, the structure factor is determined only by the temperature and the solvent viscosity, and has a universal form independent of the characteristics of the polymer. Fig. 4.4 shows the results of an experiment. Although there are no adjustable parameters in the theory, the theoretical curve agrees well with the experimental data.

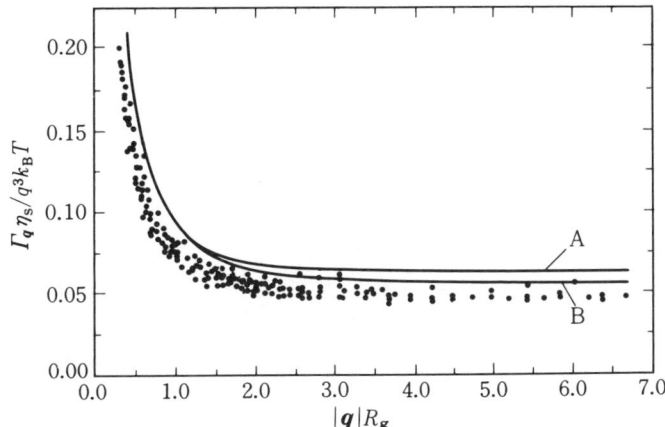

**Fig. 4.4** The relationship between the rate of decay $\Gamma_q$ and scattering vector $q$ in a dynamic light scattering experiment on a polymer solution under $\Theta$ conditions. The circles are experimental results, the solid line A is from eqn (4.110) and the line B is from the preaveraging approximation. (Han, C.C., and Akcasu, A.Z., (1981). *Macromolecules*, **14**, 1080, Fig. 5.)

For $qR_g \ll 1$, the general form of the initial decay rate can be found. Using (4.67) and (4.106), we have, for $q \to 0$,

$$\frac{d}{dt}g(q, t)|_{t=0} = -\frac{k_B T}{N}\sum_{n,m}\langle qq : H_{nm}\rangle$$

$$= -\frac{k_B T q^2}{6\pi\eta_s N}\sum_{n,m}\left\langle\frac{1}{|R_n - R_m|}\right\rangle. \tag{4.114}$$

As $q \to 0$, $g(q, 0) \to N$, giving us

$$\Gamma_q = \frac{k_B T q^2}{6\pi\eta_s N^2}\sum_{n,m}\left\langle\frac{1}{|R_n - R_m|}\right\rangle. \tag{4.115}$$

On the other hand, from (4.101), if $qR_g \ll 1$, $\Gamma_q$ is given by $D_G q^2$, and so the diffusion constant is given by

$$D_G = \frac{k_B T}{6\pi\eta_s N^2}\sum_{n,m}\left\langle\frac{1}{|R_n - R_m|}\right\rangle. \tag{4.116}$$

This equation was first found by Kirkwood, and it applies not only to flexible polymers, but also to rod-like or branched polymers. Kirkwood's equation is rigorous for the diffusion constant defined from the initial rate of decay, but is different from the long time diffusion constant. However, in most cases the Kirkwood equation is a very good approximation. Further, the diffusion constant calculated from Zimm theory agrees completely with that calculated from Kirkwood's equation.

## 4.5 Appendix

### 4.5.1 Derivation of the diffusion equation

Let us assume that $V(x, t)$ is a given function of $x, t$, and consider the following time evolution equation:

$$x(t + \Delta t) = x(t) + V(x, t)\Delta t + \Delta G(t). \tag{4.117}$$

Here $\Delta G(t)$ is a Gaussian distribution probability function with a mean and variance given by (4.15). Letting $\xi(t) \equiv V(x, t)\Delta t + \Delta G(t)$, then $\xi(t)$ also has a Gaussian distribution, with the following mean and variance:

$$\langle \xi(t) \rangle = V(x, t)\Delta t \tag{4.118}$$

$$\left\langle (\xi(t) - \langle \xi(t) \rangle)(\xi(t') - \langle \xi(t') \rangle) \right\rangle = 2D(x)\delta_{tt'}\Delta t + O(\Delta t^2). \tag{4.119}$$

(In Section 4.1, $D$ was assumed to be constant, but here for generality we will assume that $D$ is a function of $x$.) From (4.118) and (4.119), the distribution function of $\xi(t)$ is given by

$$F(\xi; x, t) = (4\pi D(x)\Delta t)^{-1/2} \exp\left[ -\frac{(\xi - V(x, t)\Delta t)^2}{4D(x)\Delta t} \right]. \tag{4.120}$$

So $F(\xi; x, t)$ is the probability that a particle located at $x$ at time $t$ will be found at $x + \xi$ at time $t + \Delta t$. With this probability function, the time evolution of $\psi(x, t)$, the distribution of $x$, is given by the following:

$$\psi(x, t + \Delta t) = \int_{-\infty}^{\infty} d\xi\, \psi(x - \xi, t)F(\xi; x - \xi, t). \tag{4.121}$$

In this equation we have used the fact that $F(\xi; x - \xi, t)$ is the probability that a particle at $x - \xi$ at time $t$ will be found at position $x$ at time $t + \Delta t$.

Since $\psi(x, t), F(\xi; x, t)$ are slowly varying functions of $x$, when we calculate the integral on the right-hand side of (4.121) for small $\xi$, the functions evaluated at $x - \xi$ can be calculated by expanding with respect to $\xi$:

$$\psi(x, t + \Delta t) = \int_{-\infty}^{\infty} d\xi \left[ 1 - \xi\frac{\partial}{\partial x} + \frac{1}{2}\xi^2\frac{\partial^2}{\partial x^2} \right] \psi(x, t)F(\xi; x, t)$$

$$= \psi(x, t) - \frac{\partial}{\partial x}\langle \xi(t) \rangle \psi(x, t) + \frac{1}{2}\frac{\partial^2}{\partial x^2}\langle \xi(t)^2 \rangle \psi(x, t). \tag{4.122}$$

Using (4.119), and noting that $\langle \xi(t)^2 \rangle = 2D(x)\Delta t + O(\Delta t^2)$ we have

$$\psi(x, t + \Delta t) - \psi(x, t) = -\frac{\partial}{\partial x}[V(x, t)\psi(x, t))]\Delta t + \frac{\partial^2}{\partial x^2}[D(x)\psi(x, t)]\Delta t. \quad (4.123)$$

Therefore

$$\frac{\partial \psi}{\partial t} = -\frac{\partial}{\partial x}[V(x, t)\psi(x, t)] + \frac{\partial^2}{\partial x^2}[D(x)\psi(x, t)]. \quad (4.124)$$

If $D$ is constant, this agrees with (4.16).

If $D > 0$, it can be shown that (4.124) approaches the following steady state solution as $t \to \infty$:

$$\psi_{\text{eq}} \propto \exp(-W(x)). \quad (4.125)$$

Here $W$ satisfies the following equation:

$$\frac{\partial W}{\partial x} = \frac{1}{D}\left(-V + \frac{\partial D}{\partial x}\right). \quad (4.126)$$

In particular, if $D$ is constant and $V = -(1/\zeta)\partial U/\partial x$, eqn (4.125) agrees with (4.17).

### 4.5.2 Initial decay rate of the time correlation function

Let us consider the time correlation function $\langle A(t)A(0) \rangle$ of the physical quantity $A$ in the equilibrium state, assuming that the distribution of the variable $x$ follows (4.19). If $A$ is a state variable and $x$ is the state of the system, the value that $A$ takes is uniquely determined and is written as $\tilde{A}(x)$. If we write $G(x, t; x', t')$ for the conditional probability that a system in state $x'$ at time $t'$ will be in a state $x$ at time $t$, then $\langle A(t)A(0) \rangle$ can be written as follows:

$$\langle A(t)A(0) \rangle = \int \mathrm{d}x \mathrm{d}x' \tilde{A}(x)G(x, t; x', 0)\tilde{A}(x')\psi_{\text{eq}}(x'). \quad (4.127)$$

Here $\psi_{\text{eq}}(x)$ is the equilibrium distribution function:

$$\psi_{\text{eq}}(x) \propto \exp\left(-\frac{U(x)}{k_B T}\right). \quad (4.128)$$

The meaning of (4.127) should be clear. The function $G(x, t; x', 0)\psi_{\text{eq}}(x')$ represents the probability that an equilibrium system in the state $x'$ at time 0 is in the state $x$ at time $t$. The physical quantity $A$ takes the values $\tilde{A}(x')$ and $\tilde{A}(x)$ at these states. Thus averaging these over $x$ and $x'$ gives us (4.127).

Further, from the definition, $G(x, t; x', 0)$ at $t = 0$ satisfies

$$G(x, 0; x', 0) = \delta(x - x'), \quad (4.129)$$

and for $t > 0$ satisfies the following time evolution equation:

$$\frac{\partial G}{\partial t} = \frac{\partial}{\partial x} D \left[ \frac{\partial G}{\partial x} + \frac{1}{k_{\mathrm{B}} T} \frac{\partial U}{\partial x} G \right]. \tag{4.130}$$

Therefore, taking the time derivative of (4.127),

$$\begin{aligned}
\frac{\mathrm{d} \langle A(t) A(0) \rangle}{\mathrm{d}t} &= \int \mathrm{d}x \mathrm{d}x' \tilde{A}(x) \frac{\partial G(x, t; x', 0)}{\partial t} \tilde{A}(x') \psi_{\mathrm{eq}}(x') \\
&= \int \mathrm{d}x \mathrm{d}x' \tilde{A}(x) \left[ \frac{\partial}{\partial x} D \left( \frac{\partial G}{\partial x} + \frac{1}{k_{\mathrm{B}} T} \frac{\partial U}{\partial x} G \right) \right] \tilde{A}(x') \psi_{\mathrm{eq}}(x') \\
&= - \int \mathrm{d}x \mathrm{d}x' \frac{\partial \tilde{A}(x)}{\partial x} D \left[ \frac{\partial G}{\partial x} + \frac{1}{k_{\mathrm{B}} T} \frac{\partial U}{\partial x} G \right] \tilde{A}(x') \psi_{\mathrm{eq}}(x'). \tag{4.131}
\end{aligned}$$

At $t = 0$, using (4.129) yields

$$\begin{aligned}
\frac{\mathrm{d} \langle A(t) A(0) \rangle}{\mathrm{d}t} \bigg|_{t=0} &= - \int \mathrm{d}x \frac{\partial \tilde{A}(x)}{\partial x} D \left[ \frac{\partial}{\partial x} + \frac{1}{k_{\mathrm{B}} T} \frac{\partial U}{\partial x} \right] \tilde{A}(x) \psi_{\mathrm{eq}}(x) \\
&= - \int \mathrm{d}x \frac{\partial \tilde{A}(x)}{\partial x} D \frac{\partial \tilde{A}(x)}{\partial x} \psi_{\mathrm{eq}}(x) \\
&= - \left\langle \frac{\partial \tilde{A}(x)}{\partial x} D \frac{\partial \tilde{A}(x)}{\partial x} \right\rangle. \tag{4.132}
\end{aligned}$$

The above is only for a single variable. For multiple variables we will have:

$$\frac{\mathrm{d} \langle A(t) A(0) \rangle}{\mathrm{d}t} \bigg|_{t=0} = - \sum_{i,j} \left\langle \frac{\partial \tilde{A}(\{x\})}{\partial x_i} D_{ij} \frac{\partial \tilde{A}(\{x\})}{\partial x_j} \right\rangle \tag{4.133}$$

$$= - \sum_{i,j} k_{\mathrm{B}} T \left\langle \frac{\partial \tilde{A}(\{x\})}{\partial x_i} \mu_{ij} \frac{\partial \tilde{A}(\{x\})}{\partial x_j} \right\rangle. \tag{4.134}$$

This equation is used in Section 4.4.2.

# 5

# Molecular motion in entangled polymer systems

If the concentration in a polymer solution exceeds $c^*$, the polymer molecules will begin to overlap. In this state, excluded volume interactions, hydrodynamic interactions, and entanglement interactions all strongly affect the molecular motion, and the calculations become extremely complicated. A rigorous treatment of the entanglement interactions is particularly difficult, and an analysis from first principles is almost impossible. For these reasons, present theories of the dynamics of concentrated polymer systems are based on very rough models, which manage to capture some features of the motion very well, but cannot describe all aspects of the dynamics. Present theories of the dynamics of entangled polymer systems are based on two theories that have met with reasonable success.

One theory describes the concentration fluctuations in concentrated polymer solutions. Here, entanglements between the polymers are ignored, and the combined movements of the segments are considered. Another theory is the reptation model, which describes viscoelasticity and diffusion in concentrated solutions and melts. In this model, the motion of the individual polymers is considered, where the entanglement effects are very important, but the combined motion of the polymers is neglected.

These two theories describe different aspects of the motion of polymer chains. However, it now seems that the two types of motion considered by these theories are not independent but are actually related, and this is supported by several experimental results. However, at present this relationship is not fully understood, so in this chapter we will regard them as two separate theories.

## 5.1 Dynamics of concentration fluctuations

### 5.1.1 Time correlation function of concentration fluctuations

The concentration fluctuations in a concentrated polymer system can be investigated by dynamic light scattering. We will label the polymers 1,2,...,

$n_p$, and write the position vector of the $n$th segment of the $a$th polymer as $\boldsymbol{R}_{an}(t)$. At time $t$ and at location $\boldsymbol{r}$ the segment concentration is

$$c(\boldsymbol{r}, t) = \sum_{a=1}^{n_p} \sum_{n=1}^{N} \delta(\boldsymbol{r} - \boldsymbol{R}_{an}(t)). \tag{5.1}$$

The Fourier transform of this is

$$c_{\boldsymbol{q}}(t) = \int \mathrm{d}\boldsymbol{r} e^{i\boldsymbol{q}\cdot\boldsymbol{r}} c(\boldsymbol{r}, t) = \sum_{a=1}^{n_p} \sum_{n=1}^{N} \exp[i\boldsymbol{q} \cdot \boldsymbol{R}_{an}(t)]. \tag{5.2}$$

The dynamic structure factor $g(\boldsymbol{q}, t)$ can be expressed in terms of the time correlation function of $c_{\boldsymbol{q}}(t)$ as follows:

$$g(\boldsymbol{q}, t) = \frac{1}{cV} \left\langle c_{\boldsymbol{q}}(t) c_{-\boldsymbol{q}}(0) \right\rangle. \tag{5.3}$$

Here $V$ is the volume of the system and $c$ is the average segment concentration ($c = n_p N/V$). Using (5.2) we have

$$g(\boldsymbol{q}, t) = \frac{1}{n_p N} \sum_{a,b=1}^{n_p} \sum_{n,m=1}^{N} \left\langle \exp[i\boldsymbol{q} \cdot (\boldsymbol{R}_{an}(t) - \boldsymbol{R}_{bm}(0))] \right\rangle. \tag{5.4}$$

For dilute solutions, (5.4) agrees with eqn (4.99).

### 5.1.2 Cooperative diffusion constant

For dilute solutions, the diffusion constant of the individual polymers can be determined from the decay rate $\Gamma_{\boldsymbol{q}}$ of $g(\boldsymbol{q}, t)$. However, this becomes impossible for concentrations where the polymers begin to overlap, because the dynamic structure factor in concentrated solutions represents the concentration fluctuations of segments belonging to many different polymers, and does not reflect the motion of individual polymers. For concentrated solutions, the motion of the individual polymers would be expected to be slowed down due to entanglement effects, but in fact, $g(\boldsymbol{q}, t)$ is found to decay even faster. The reason is that for high polymer concentrations, the excluded volume effect increasingly acts to make the concentration uniform as fast as possible. Let us now estimate this effect using a simple molecular field theory.

Let us consider a segment at position $\boldsymbol{r}$. The mean field experienced by this segment can be written in terms of the segment concentration $c(\boldsymbol{r})$ and the excluded volume parameter $v$ as follows:

$$w(\boldsymbol{r}) = v k_{\mathrm{B}} T c(\boldsymbol{r}). \tag{5.5}$$

If the segment concentration is not uniform, there is a force $-\nabla w$ which acts on each segment, causing it to move with the following average velocity:

$$V(r) = -\frac{1}{\zeta}\nabla w(r) = -\frac{v}{\zeta}k_B T \nabla c(r, t). \tag{5.6}$$

Here $\zeta$ is the friction constant of the segment. Substituting (5.6) into the equation of continuity

$$\frac{\partial}{\partial t}c = -\nabla \cdot Vc. \tag{5.7}$$

and linearizing with respect to the concentration fluctuation $\delta c(r, t) = c(r, t) - \bar{c}$, gives us

$$\frac{\partial}{\partial t}\delta c = \frac{v\bar{c}}{\zeta}k_B T \nabla^2 \delta c(r, t)$$
$$= D_c \nabla^2 \delta c(r, t). \tag{5.8}$$

Here

$$D_c = \frac{v\bar{c}}{\zeta}k_B T \tag{5.9}$$

is called the co-operative diffusion constant. From (5.8), $c_q$ becomes

$$c_q(t) = c_q(0)\exp(-D_c q^2 t). \tag{5.10}$$

Therefore, the dynamic structure factor becomes

$$g(q, t) = g(q, 0)\exp(-D_c q^2 t). \tag{5.11}$$

Equation (5.11) has the same form as eqn (4.101), the dynamic structure factor for dilute solutions, except that $D_G$ is replaced by the co-operative diffusion constant $D_c$.

The diffusion constant in (4.101) is called the self-diffusion constant. When calculating this, we focus on one polymer and consider the distance moved by the centre of mass $R_G(t)$ during time $t$. We can define the self-diffusion constant from the average squared displacement $\langle (R_G(t) - R_G(0))^2 \rangle$ as follows:

$$D_G = \lim_{t \to \infty}\frac{1}{6t}\langle (R_G(t) - R_G(0))^2 \rangle. \tag{5.12}$$

For dilute solutions, the self-diffusion constant can be measured by dynamic light scattering, but for concentrated solutions we must rely on techniques that label the polymers, such as NMR or forced Rayleigh scattering.

The co-operative diffusion constant reflects the speed at which non-uniformities in the segment concentration are propagated through the system. For example, if we assume that for some reason there is a change in the segment concentration in a particular location, then after a time $t$ the effects of this change will be propagated through a distance $l \simeq (D_c t)^{1/2}$. This propagation is due to the excluded volume effect, and does not mean that each segment moves a distance $l$.

In Fig. 5.1, the concentration dependence of the self- and co-operative diffusion constants in polymer solutions are plotted. The co-operative diffusion constant increases with concentration, while the self-diffusion constant decreases. This is because as the polymer concentration is increased, the relaxation of the concentration fluctuations is more rapid, but the diffusion of the individual polymers is hindered by the entanglement effects.

### 5.1.3 Initial decay rate of $g(\boldsymbol{q},t)$

In the following, we will ignore the connectivity between the segments as well as the hydrodynamic interactions. The calculation of $g(\boldsymbol{q}, t)$, taking into account these effects, is very difficult. Here we will only consider $\Gamma_q$, the initial decay rate of $g(\boldsymbol{q}, t)$.

As can be seen from (4.110), if we know the static structure factor $g(\boldsymbol{q})$ then we can calculate $\Gamma_q$. Therefore, even for concentrated solutions, the initial rate of decay is given by (4.111), the same as for dilute solutions. In a

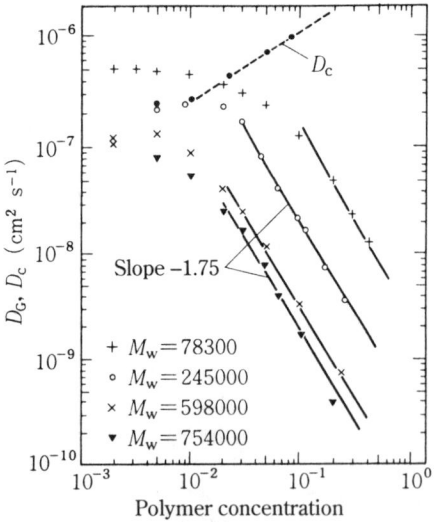

**Fig. 5.1** The self-diffusion constant $D_G$ and cooperative diffusion constant $D_c$ for polystyrene polymers in benzene. (Hervet, H., Leger, L., and Rondelez, F. (1979). *Phys. Rev. Lett.*, **42**, 1681, Fig. 2.)

concentrated system, $\xi$ is small, and so for the normal range of light scattering, we have $q\xi \ll 1$. Therefore, $\Gamma_q$ is given by (4.113):

$$\Gamma_q = \frac{k_B T}{6\pi\eta_s\xi} q^2. \tag{5.13}$$

Thus, we have

$$D_c = \frac{k_B T}{6\pi\eta_s\xi}. \tag{5.14}$$

From (2.82), we have $\xi \propto c^{-3/4}$, and so

$$D_c \propto c^{3/4}. \tag{5.15}$$

The concentration dependence of (5.15) differs from (5.9). However, (5.15) is closer to the results obtained from experiments (Fig. 5.1).

Equation (5.14) shows that the co-operative diffusion constant $D_c$ is the self-diffusion constant of a sphere of radius $\xi$. This is because if a concentration fluctuation occurs in a concentrated system, segments within a distance of the correlation length $\xi$ move almost as a solid body.

From experiments, it is known that $g(q, t)$ for a good solvent has approximately a single relaxation, with the relaxation speed close to that given by (5.13). On the other hand, in a poor solvent there appear long-time relaxation modes in $g(q, t)$, and the behaviour seems to be quite complicated.

## 5.2 Reptation

### 5.2.1 Entanglement effects and the tube model

The arguments of the previous section completely ignored entanglement effects. Despite this, we obtained results showing good agreement with experiments. The reason for this is that in co-operative diffusion, the segments move as a whole. However, when we consider the self-diffusion of the polymers, the entanglement effects become very important. If we fix our attention on a single polymer in a concentrated solution, we see that it diffuses through a network made by the surrounding polymers. Therefore the self-diffusion constant of a polymer becomes very small due to entanglement effects.

The calculation of the self-diffusion constant in concentrated solutions met with very little success for a long time due to the difficulty of treating the entanglement effects theoretically. The application of the tube concept of de Gennes in 1971 provided the key for solving this problem.

De Gennes thought of the problem in the following way. Assume that a polymer molecule is undergoing Brownian motion in a fixed network, as in Fig. 5.2(a). The dots in the figure represent the network, and we assume that the polymer cannot cut through these. We will use the Rouse model to describe the motion of the polymer, and so the self-diffusion constant of the polymer should be expressible in terms of the parameters characterizing the Rouse model, which are $N$ (the number of segments), $b$ (the segment length), and $\zeta$ (the friction constant of the segment), as well as the parameter characterizing the network $a$ (the average mesh size). What form will the diffusion constant take?

It is difficult to solve this problem rigorously. De Gennes thus used the tube concept introduced by Edwards and argued as follows. Since the polymer is not allowed to cut through the network points, we can assume that the polymer effectively undergoes motion in a tube as in Fig. 5.2(b). The diameter of this tube is of order $a$. In Chapter 3, we considered a model where both ends of the polymer are fixed to the extremities of the tube, but in the current case the polymer is free to diffuse in the tube. Because of this, as time passes, the polymer is able to wriggle out of the tube in which it initially found itself, as in Fig. 5.2(c). However, this is not to say that the constraints on the polymer due to the network have vanished, but rather that the protruding portion CD moves into a newly created tube, and the part of the tube AB vacated by the polymer disappears. In other words, the polymer uses the degree of freedom at its ends to change its shape gradually during its motion along itself. De Gennes called this type of motion 'reptation'.

### 5.2.2 Lattice model of reptation

Let us consider the following model of reptation motion. As shown in Fig. 5.3, let us place a polymer consisting of $Z$ bonds on a lattice of spacing $a$.

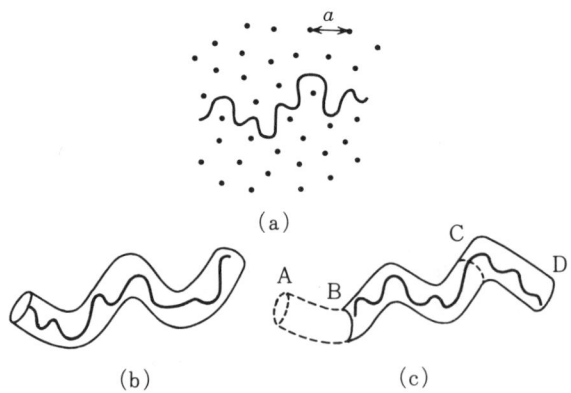

**Fig. 5.2** (a) A polymer moving in a fixed network; (b) the tube model; (c) the situation depicted in (b) after some time has passed.

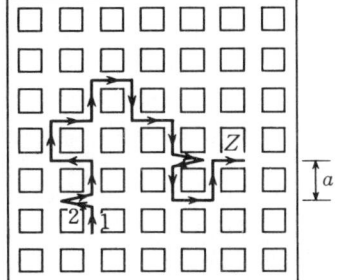

**Fig. 5.3** The lattice model of reptation.

Let the directions of the bond vectors be represented by the unit vectors $u_1, u_2, ..., u_Z$. In the equilibrium state these will point in random directions. In order to simulate the reptation motion, let us assume that the head or the tail of the polymer advances by one space during a time interval $\Delta t$. If it is the head that advances, then all the bond vectors from 2 to $Z$ change from $u_n$ to $u_{n-1}$, and $u_1$ takes an independent random direction. Similarly, if it is the tail that advances, then $u_n$ changes to $u_{n+1}$, and $u_Z$ takes a random direction. Expressing this mathematically,

$$u_n(t + \Delta t) = u_{n+\xi(t)}(t). \tag{5.16}$$

Here, $\xi(t)$ is a random variable taking the values $+1$ or $-1$, and $u_{-1}$ and $u_{Z+1}$ are unit vectors that can point in random directions.

The parameters $\Delta t$ and $Z$ appearing in this lattice model of reptation can be expressed in terms of the parameters of the Rouse model. In the lattice model the average squared end-to-end distance of the polymer is $Za^2$. On the other hand, in the Rouse model the same quantity is given by $Nb^2$. Therefore

$$Z = \frac{Nb^2}{a^2}. \tag{5.17}$$

To find $\Delta t$ we consider the diffusion of the polymer along its own length. In the Rouse model, the centre of mass will move with a diffusion constant $k_B T/N\zeta \equiv D_t$ in the absence of obstacles. Since there are no obstacles to the polymer when it moves along the tube, during a time $t$ the average of the square of the distance moved by the polymer along the tube $s(t)$ is given by $\langle s^2(t) \rangle = 2D_t t$.[1] On the other hand, in the lattice reptation model, in a time

---

[1] Strictly speaking, the motion of the polymer in the tube is affected by the presence of the tube. However, this can be incorporated into the friction constant $\zeta$, and so here we will not explicitly consider this effect.

interval $t$ the chain experiences $t/\Delta t$ random jumps, and with each jump the chain moves by a distance $a$. Thus we have $\langle s^2(t) \rangle = (t/\Delta t)a^2$. Therefore we have

$$\Delta t = \frac{a^2}{2D_t} = \frac{a^2 N \zeta}{2k_B T}. \tag{5.18}$$

The basic equation for the lattice reptation model can be written as follows using the coordinates of the segments $R_n(n = 0, 1, ..., Z)$:

$$R_n(t + \Delta t) = R_{n+\xi(t)}(t). \tag{5.19}$$

Here $R_{-1}(t + \Delta t)$ and $R_{Z+1}(t + \Delta t)$ can be defined as follows using the random unit vector $v(t)$:

$$R_{-1}(t + \Delta t) = R_0(t) - av(t), \quad R_{Z+1}(t + \Delta t) = R_Z(t) + av(t). \tag{5.20}$$

Using (5.19) and (5.20), let us investigate the features of Brownian motion of the reptation model.

### 5.2.3 Motion of the centre of mass

The centre of mass of the polymer is given by

$$R_G(t) = \frac{1}{Z+1} \sum_{n=0}^{Z} R_n(t). \tag{5.21}$$

If $\xi(t) = 1$, then using (5.19) and (5.20) we can write

$$\begin{aligned}
R_G(t + \Delta t) &= \frac{1}{Z+1} \left[ \sum_{n=1}^{Z} R_n(t) + R_Z(t) + av(t) \right] \\
&= R_G(t) + \frac{1}{Z+1} (R_Z(t) + av(t) - R_0(t)) \\
&= R_G(t) + \frac{1}{Z+1} (P(t) + av(t)).
\end{aligned} \tag{5.22}$$

Here

$$P(t) = R_Z(t) - R_0(t). \tag{5.23}$$

is the end-to-end vector. Similarly, if $\xi(t) = -1$, we have

$$R_G(t + \Delta t) = R_G(t) - \frac{1}{Z+1} (P(t) + av(t)). \tag{5.24}$$

Equations (5.22) and (5.24) can be written as follows:

$$R_G(t + \Delta t) = R_G(t) + \xi(t)f(t). \tag{5.25}$$

Here

$$f(t) = \frac{P(t) + av(t)}{Z + 1}. \tag{5.26}$$

Now, the time correlation of the second term on the right-hand side of (5.25), defined as $C_{\xi f} \equiv \langle \xi(t)f(t) \cdot \xi(t')f(t') \rangle$, is 0 for times other than $t = t'$. The reason is that if $t > t'$, then $\xi(t)$ is independent of other quantities and so $C_{\xi f} = \langle \xi(t) \rangle \langle f(t) \cdot \xi(t')f(t') \rangle = 0$. Similarly, if $t < t'$, then $C_{\xi f}$ equals 0, and so we have

$$\langle \xi(t)f(t) \cdot \xi(t')f(t') \rangle = \delta_{tt'} \langle f(t)^2 \rangle. \tag{5.27}$$

Therefore

$$\langle (R_G(t) - R_G(0))^2 \rangle = \frac{t}{\Delta t} \langle f(t)^2 \rangle. \tag{5.28}$$

On the other hand, at equilibrium $\langle P^2(t) \rangle = Za^2$ and so we have

$$\langle f(t)^2 \rangle = \frac{\langle P^2 \rangle + a^2}{(Z+1)^2} = \frac{1}{(Z+1)} a^2. \tag{5.29}$$

Using (5.28) and (5.29) we have

$$\langle (R_G(t) - R_G(0))^2 \rangle = \frac{t}{\Delta t} \frac{a^2}{Z}$$
$$= \frac{2D_t}{Z} t. \tag{5.30}$$

Here we have assumed $Z \gg 1$. Therefore, the self-diffusion constant $D_G$ becomes

$$D_G = \frac{D_t}{3Z} \tag{5.31}$$

$$= \frac{k_B T}{3N^2 \zeta} \left( \frac{a^2}{b^2} \right). \tag{5.32}$$

Thus we see that the diffusion constant of the centre of mass is proportional to $N^{-2}$, and decreases as the lattice spacing $a$ decreases.

As shown in Fig. 5.1, the self diffusion constant decreases as the polymer molecular weight $M$ increases or as the polymer concentration increases. The result $D_G \propto M^{-2}$ is found to agree with experimental measurements on polymer solutions or melts (Fig. 5.4).

*5.2.4 Rotational motion*

The correlation function $\langle P(t) \cdot P(0) \rangle$ of the end-to-end vector $P(t)$ can be expressed in terms of the correlation function of the bond vectors $u_n(t)$.

$$\psi_{n,m}(t) \equiv \langle u_n(t) \cdot u_m(0) \rangle, \tag{5.33}$$

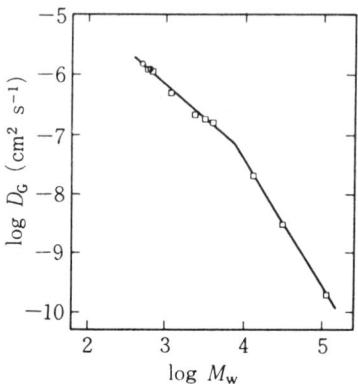

**Fig. 5.4** The self-diffusion constant of a molecule in a polyethylene melt. The two straight sections have slopes of $-1$ and $-2$. (Pearson, D.S., Verstrate, V., Meerwall, E. von, and Schilling, F.C. (1987). *Macromolecules*, **20**, 1133, Fig. 11.)

as follows:

$$\langle \boldsymbol{P}(t) \cdot \boldsymbol{P}(0) \rangle = a^2 \sum_{n=1}^{Z} \sum_{m=1}^{Z} \psi_{n,m}(t). \tag{5.34}$$

Now after a time $\Delta t$, $\boldsymbol{u}_n(t)$ changes to $\boldsymbol{u}_{n+1}(t)$ or $\boldsymbol{u}_{n-1}(t)$, and so $\psi_{n,m}(t)$ satisfies the following equation:

$$\psi_{n,m}(t + \Delta t) = \frac{1}{2}[\psi_{n+1,m}(t) + \psi_{n-1,m}(t)]. \tag{5.35}$$

Now $\boldsymbol{u}_0$ corresponds to the random vector $\boldsymbol{v}(t)$, and so

$$\psi_{0,m}(t) = \langle \boldsymbol{u}_0(t) \cdot \boldsymbol{u}_m(0) \rangle = \langle \boldsymbol{u}_0(t) \rangle \cdot \langle \boldsymbol{u}_m(0) \rangle = 0. \tag{5.36}$$

Similarly,

$$\psi_{Z+1,m}(t) = 0. \tag{5.37}$$

Further, at $t = 0$ we clearly have

$$\psi_{n,m}(0) = \delta_{nm}. \tag{5.38}$$

Equations (5.35)–(5.38) represent a set of difference equations for $\psi_{n,m}(t)$, for which a solution can be calculated. For simplicity of calculation, let us assume that $Z \gg 1$. Then $\psi_{n,m}(t)$ becomes a slowly varying function of $t$ and $n$, and so the left- and right-hand sides of (5.35) can be rewritten as follows:

$$\psi_{n,m}(t + \Delta t) = \psi_{n,m}(t) + \frac{\partial \psi_{n,m}}{\partial t} \cdot \Delta t = \psi_{n,m} + \frac{\partial \psi_{n,m}}{\partial t} \cdot \frac{a^2}{2D_t}. \tag{5.39}$$

$$\frac{1}{2}(\psi_{n+1,m}(t) + \psi_{n-1,m}(t)) = \psi_{n,m} + \frac{1}{2}\frac{\partial^2 \psi_{n,m}}{\partial n^2}. \tag{5.40}$$

Thus, (5.35) becomes the following partial differential equation:

$$\frac{\partial \psi_{n,m}}{\partial t} = \frac{D_t}{a^2} \frac{\partial^2 \psi_{n,m}}{\partial n^2}. \tag{5.41}$$

Further, eqns (5.36)–(5.38) become the following boundary and initial conditions:

$$\psi_{0,m}(t) = \psi_{Z,m}(t) = 0 \qquad \psi_{n,m}(0) = \delta(n - m). \tag{5.42}$$

Solving these equations gives us

$$\psi_{n,m}(t) = \frac{2}{Z} \sum_{p=1}^{\infty} \sin\left(\frac{np\pi}{Z}\right) \sin\left(\frac{mp\pi}{Z}\right) \exp\left(-\frac{tp^2}{\tau_d}\right). \tag{5.43}$$

Here, $\tau_d$ is the 'reptation time' defined as follows:

$$\tau_d = \frac{Z^2 a^2}{\pi^2 D_t}. \tag{5.44}$$

Substituting (5.43) into (5.34) gives

$$\langle \boldsymbol{P}(t) \cdot \boldsymbol{P}(0) \rangle = Z a^2 \psi(t). \tag{5.45}$$

Here

$$\psi(t) = \sum_p \frac{8}{p^2 \pi^2} \exp\left(-\frac{tp^2}{\tau_d}\right), \tag{5.46}$$

where $p$ ranges over positive odd integers. We see that $\psi(0) = 1$, and that as $t$ increases, $\psi(t)$ decreases with a relaxation time $\tau_d$. Therefore, the rotational relaxation time is given by the reptation time $\tau_d$. Using (5.17), we have

$$\tau_d = \frac{1}{\pi^2} \frac{\zeta N^3 b^4}{k_B T a^2}. \tag{5.47}$$

We see that $\tau_d$ is proportional to $N^3$.

As we shall see later, $\tau_d$ can be found from elasticity experiments. From these measurements, it is found that $\tau_d$ is not proportional to $N^3$ but to a slightly higher exponent of $N$, between 3.3 and 3.5 (Fig. 5.5). This discrepancy can be explained by considering the fluctuations in the tube length (see Section 5.2.6).

### 5.2.5 Shape memory function

Let us consider the physical meaning of the function $\psi(t)$ defined in (5.45). Due to the reptation motion of the polymer, the ends of the polymer gradually wriggle out of the tubes in which they initially were. For example, assume that the polymer moves as in Fig. 5.6. The section CD in Fig. 5.6(d) is inside the tube of time $t = 0$, but the sections AC and DB move into a new

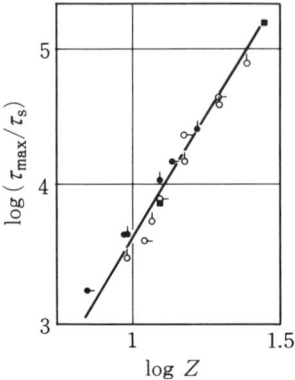

**Fig. 5.5** The relationship between the longest relaxation time $\tau_{max}$ and $Z$ in a polymer solution. The quantity $\tau_s$ is a characteristic relaxation time obtained from the small time components of the elastic relaxation modulus. Squares: polystyrene melt; black circles: polystryene–alcohol solution; white circles: poly($\alpha$-methyl styrene)–alcohol solution. The slope of the line in the figure is 3.4. (Osaki, K. (1988). In *Molecular conformation and dynamics of macromolecules in condensed systems*, (ed. M. Nagasawa), p. 189, Fig. 2.3. Elsevier, Amsterdam.)

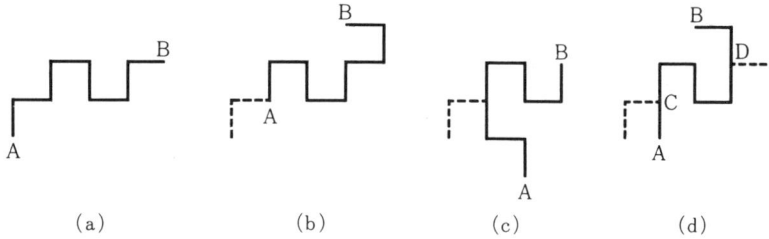

**Fig. 5.6** An example of the changes in the polymer configuration with the lattice model of reptation. The polymer initially in the state shown in (a) moves 2 steps in the B direction to give the configuration in (b). It then moves 3 steps in the A direction followed by 2 steps in the B direction to give (c) and (d).

tube. In the state (d), $\overrightarrow{AC}$ and $\overrightarrow{DB}$ are made up of random vectors which are independent of the previous state, and so $\langle \overrightarrow{AC} \cdot \overrightarrow{CD} \rangle = 0$, $\langle \overrightarrow{DB} \cdot \overrightarrow{CD} \rangle = 0$. Therefore, $\langle \boldsymbol{P}(t) \cdot \boldsymbol{P}(0) \rangle = \langle \overrightarrow{CD}^2 \rangle$. Thus, writing $l(t)$ for the number of bonds in the section of the original tube of $t = 0$ still remaining, the correlation function $\langle \boldsymbol{P}(t) \cdot \boldsymbol{P}(0) \rangle$ can be written as

$$\langle \boldsymbol{P}(t) \cdot \boldsymbol{P}(0) \rangle = a^2 \langle l(t) \rangle. \tag{5.48}$$

Therefore, from (5.45)

$$\psi(t) = \frac{\langle l(t) \rangle}{Z}. \tag{5.49}$$

In other words, $\psi(t)$ is the fraction of the polymer at time $t$ remaining in the original tube of $t = 0$.

Since the polymer moves along the tube with a diffusion constant $D_t$, in a time interval $t$ it moves a distance of order $\sqrt{D_t t}$ along the tube. If this distance is of the same order as the length of the polymer $L = Za$, then the polymer completely escapes from the tube. Therefore, the time for escape from the tube can be estimated as $Z^2 a^2 / D_t$. Ignoring numerical coefficients, we see that this time is the same as the reptation time of (5.44). Therefore, the reptation time is also called the escape time.

Using the same arguments, the diffusion constant of the centre of mass can also be estimated. After a time $\tau_d$, the polymer leaves the tube it was in and enters a new tube. In this time, the centre of mass of the polymer moves a distance of the order of the radius of gyration $R_g \simeq \sqrt{N} b$. During a time $t$, this process is repeated $t/\tau_d$ times, and so the average squared distance is $(t/\tau_d) N b^2$. Thus, the diffusion constant of the centre of mass is given by $N b^2 / \tau_d$. This agrees with (5.32).

### 5.2.6 Fluctuations of the tube contour length

In the lattice model of reptation, we have assumed that the length of the polymer along the tube $L$ is fixed. However, we would expect this length to fluctuate in reality. Let us write $\Psi(L)$ for the probability distribution of the length $L$. Then $\Psi(L)$ is proportional to the product of the number of states $W_p(L)$ that a polymer trapped inside a tube of length $L$ can take, and the number of states $W_t(L)$ that a tube of length $L$ can take (i.e. the number of ways of placing a tube of length $L$ on the lattice of Fig. 5.3). Now $W_p(L)$ can be estimated by (3.59) in Section 3.3.3, and if we write $z$ for the coordination number of the lattice, $W_t(L)$ can estimated by $z^{(L/a)}$. Therefore, we have

$$\Psi(L) \propto W_p(L) W_t(L) \propto \exp\left(-\frac{3L^2}{2Nb^2} + \frac{L}{a} \ln z\right). \tag{5.50}$$

From (5.50), the maximum of $\Psi(L)$ occurs at

$$L^* = \frac{\ln(z)}{3} \frac{Nb^2}{a}. \tag{5.51}$$

This differs from the result we have been using up to now, $L = Za = Nb^2/a$, by a numerical factor of $\ln z/3$, but we need not be too concerned by this difference at the current rough level of approximation. Thus, in the following calculations we will write the distribution function $\Psi(L)$ as follows:

$$\Psi(L) \propto \exp\left[-\frac{3}{2Nb^2}(L - \overline{L})^2\right].$$ (5.52)

Here

$$\overline{L} = \frac{Nb^2}{a}.$$ (5.53)

Thus, the fluctuations of $L$ are given by

$$\langle(L - \overline{L})^2\rangle = \frac{Nb^2}{3}.$$ (5.54)

Therefore, the ratio of the fluctuations to the average value $\overline{L}$ is given by

$$\frac{\langle\Delta L^2\rangle^{1/2}}{\overline{L}} \cong \frac{1}{\sqrt{Z}}.$$ (5.55)

As shown in Fig. 5.7, the extensional and contractual motion of the polymer along the length of the tube can be thought of as the linear motion of a Rouse chain trapped inside a tube. Thus, the characteristic time of this motion is the same as the relaxation time of the Rouse model:

$$\tau_R = \frac{\zeta N^2 b^2}{3\pi^2 k_B T}.$$ (5.56)

Thus, the ratio of the extensional characteristic time to the reptation time is

$$\frac{\tau_R}{\tau_d} = \frac{a^2}{3Nb^2} = \frac{1}{3Z}.$$ (5.57)

According to (5.55) and (5.57), if $Z \gg 1$ we see that for slow overall motions of the polymer, the length fluctuations due to extensional or contractual motions can be ignored. However, for usual polymers, $Z$ is at the most of the order of 100. In this case extensional motions can have a non-negligible effect on the reptation time.

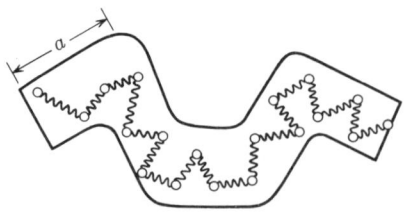

**Fig. 5.7** A Rouse chain trapped in a tube.

Assume that the ends of the polymer are rapidly fluctuating with a range of $\Delta L \simeq \langle \Delta L^2 \rangle^{1/2}$. Thus, if the polymer moves a distance of $L - \Delta L$, it can escape from the tube it was originally in. Therefore, the time required for escape, $\tau_d$, is given by $(L - \Delta L)^2/D_t$, and not by $L^2/D_t$. Using (5.55), we have

$$\tau_d \propto Z^3 \left( 1 - \frac{C}{\sqrt{Z}} \right)^2. \tag{5.58}$$

Here, $C$ is a numerical constant. The result of detailed calculations gives $C$ as approximately 1.5. For $10 < Z < 100$, (5.58) is very close to $\tau_d \propto Z^{3.4}$. Thus the dependence of the reptation time on the molecular weight can be explained if we consider the effect of the extensional motion.

## 5.3 Viscoelasticity of polymers

### 5.3.1 Phenomenological theory of viscoelasticity

A polymeric liquid, whilst retaining the properties of a liquid, also shows a rubber-like elasticity. An example is the melted cheese on a pizza. If melted cheese is dripped vertically, it flows slowly, just as a liquid. However, if it is strongly pulled and then the tension removed, melted cheese will contract just like rubber. In other words, although melted cheese is a liquid, it also has elasticity. Substances like this, which have both viscous and elastic properties, are called viscoelastic. Viscoelasticity is a characteristic of polymer systems, and can be observed in almost all materials containing polymers.

If we try to calculate the flow of a fluid when an external force is applied, we need an equation relating the stress in the fluid to its deformation. This type of equation is called a constitutive equation.

Generally, a constitutive equation relates the stress tensor $\sigma_{\alpha\beta}$ to the velocity gradient tensor $\kappa_{\alpha\beta} = \partial v_\alpha/\partial r_\beta$. For normal viscous fluids, the relationship takes the following form, writing $\eta$ for the viscosity of the fluid:

$$\sigma_{\alpha\beta} = \eta(\kappa_{\alpha\beta} + \kappa_{\beta\alpha}) - P\delta_{\alpha\beta}. \tag{5.59}$$

The constitutive equation for polymeric liquids cannot be written in such a simple form. As an example, let us consider the following shear flow:

$$v_x = \dot{\gamma}(t)y, \qquad v_y = 0 \qquad v_z = 0. \tag{5.60}$$

Here $\dot{\gamma} \equiv \partial v_x/\partial y$ is called the shear rate. In the case of a viscous liquid, (5.59) gives the following for the stress:

$$\sigma_{xy}(t) = \sigma_{yx} = \eta\dot{\gamma}(t), \qquad \sigma_{xx} = \sigma_{yy} = \sigma_{zz}, \qquad \text{other off-diagonal components} = 0. \tag{5.61}$$

In other words, the stress at a time $t$ is determined only by the shear rate at that particular instant, and there is a proportionality relationship between the two quantities. However, for polymeric liquids the stress depends on the previous shear rates as well, and furthermore this relationship is non-linear. For example, if a polymeric liquid undergoing steady flow is stopped, the stress does not immediately become 0, but decays with a relaxation time $\tau$. Here $\tau$ depends strongly on the molecular weight of the polymer and the temperature, and can be of the order of several minutes to hours in some cases. Further, the shear stress $\sigma_{xy}$ under constant shear rate is not proportional to the shear rate $\dot{\gamma}$, and so the ratio of the two, $\sigma_{xy}/\dot{\gamma}$ (this is normally called the viscosity), can change by several orders of magnitude (Fig. 5.8). In general, this relationship between the stress and shear rate for polymeric liquids is very complicated. Industrially, this is a very important problem and has been the subject of much research, but we will not enter into the details here. The interested reader should consult the references listed at the end of this book.

If the stress is small, we can assume that there is a linear relationship between the stress and the rate of deformation. In the case of shear flow, this relationship takes the following form:

$$\sigma_{xy}(t) = \int_{-\infty}^{t} dt' G(t - t') \dot{\gamma}(t'). \tag{5.62}$$

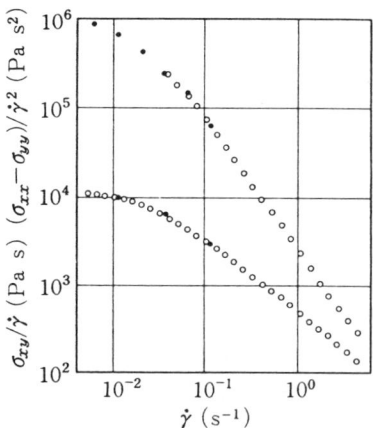

**Fig. 5.8** The variation of viscosity $\sigma_{xy}/\dot{\gamma}$ and the first normal stress difference coefficient $(\sigma_{xx} - \sigma_{yy})/\dot{\gamma}^2$ with shear rate $\dot{\gamma}$ for a polystyrene–alcohol solution under constant shear flow. The white circles are the results of direct stress measurements and the black circles are calculated from birefringence measurements using the stress optical law. (Takahashi, M., Masuda, T., Bessho, N., and Osaki, K. (1980). *J. Rheology*, **24**, 517, Fig. 2.)

Here $G(t)$ is called the stress relaxation function. In the range where this type of relationship holds, we say the system is in the regime of 'linear viscoelasticity'. In the region of linear viscoelasticity, the stress response to an arbitrary deformation can be calculated. For example, assume that we start a shear flow from time 0, with a constant shear rate $\dot{\gamma}$. From (5.62), the shear stress increases and reaches a steady value determined by

$$\sigma(t) = \dot{\gamma} \int_0^t dt' G(t - t') = \dot{\gamma} \int_0^t dt' G(t'). \qquad (5.63)$$

Thus the steady viscosity, which is the ratio of the shear stress to the shear rate at steady state, is given by

$$\eta = \int_0^\infty dt G(t). \qquad (5.64)$$

*5.3.2 Microscopic description of stress*

Let us now discuss viscoelasticity in terms of the molecular model presented in Section 5.2. In general, to calculate the viscoelasticity based on a microscopic model, it is necessary to know two things:

(1) what is the equation of motion of the polymer in a flow field?
(2) how do we calculate the macroscopic stress of the system?

The answer to (1) will be discussed in the next section. Here we will consider (2).

To generalize the discussion, let us consider a collection of mutually interacting particles suspended in a viscous fluid. Writing $\boldsymbol{R}_n$ for the position of the $n$th particle, we will assume that the interaction between particles $n$ and $m$ can be represented by the potential $u_{nm}(\boldsymbol{R}_n - \boldsymbol{R}_m)$. To calculate the stress $\sigma_{\alpha\beta}$, let us consider a planar slice of area $S$ perpendicular to the $\beta$-axis in a fluid of volume $V$ at a height $h$, as shown in Fig. 5.9. The component of the stress tensor $\sigma_{\alpha\beta}$ is the $\alpha$-component of the force exerted by the particles above the plane on the particles under the plane, divided by the area $S$.

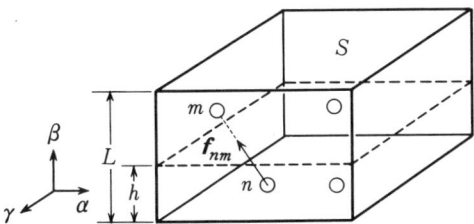

**Fig. 5.9** The stress in a suspension of interacting particles.

Therefore, writing $f_{nm} = -\partial u_{nm}(R_n - R_m)/\partial R_n$ for the force between particles $n$ and $m$, we have

$$\sigma_{\alpha\beta} = \frac{1}{S}\sum_{n,m} f_{nm\alpha}\;\Theta(h - R_{n\beta})\;\Theta(R_{m\beta} - h) + \eta_s(\kappa_{\alpha\beta} + \kappa_{\beta\alpha}) - P\delta_{\alpha\beta}. \tag{5.65}$$

Here, the two $\Theta$ step functions take into account the fact that the only contributions to the stress are when particle $m$ is above the plane and particle $n$ is below the plane. Further, the last two terms are the contributions from the forces transmitted through the solvent, but for concentrated polymer solutions, the solvent viscosity term $\eta_s(\kappa_{\alpha\beta} + \kappa_{\beta\alpha})$ is very small, and we will ignore this term from now on.

Now, in a uniform flow field, $\sigma_{\alpha\beta}$ should be independent of $h$, and so we can take an average over $h$ as follows:

$$\sigma_{\alpha\beta} = \frac{1}{L}\int_0^L dh\sigma_{\alpha\beta} - P\delta_{\alpha\beta}. \tag{5.66}$$

Here, $L$ is the height of the fluid region under consideration, and $V = SL$. Substituting into (5.65) and using the following relationship

$$\int_0^L dh\,\Theta(h - R_{n\beta})\;\Theta(R_{m\beta} - h) = R_{mn\beta}\;\Theta(R_{mn\beta}) \quad (\text{here } R_{mn} = R_m - R_n). \tag{5.67}$$

we can write (5.66) in the following way:

$$\begin{aligned}
\sigma_{\alpha\beta} + P\delta_{\alpha\beta} &= \frac{1}{V}\sum_{n,m} f_{nm\alpha}\;R_{mn\beta}\;\Theta(R_{mn\beta}) \\
&= \frac{1}{2V}\sum_{n,m}[f_{nm\alpha}\;R_{mn\beta}\;\Theta(R_{mn\beta}) - f_{nm\alpha}\;R_{nm\beta}\;\Theta(R_{nm\beta})] \\
&= -\frac{1}{2V}\sum_{n,m} f_{nm\alpha}\;R_{nm\beta} = -\frac{1}{V}\sum_{n<m} f_{nm\alpha}\;R_{nm\beta}.
\end{aligned} \tag{5.68}$$

Here we have used $f_{nm} = -f_{mn}$, $R_{nm} = -R_{mn}$, and $\Theta(R_{nm\beta}) + \Theta(R_{mn\beta}) = 1$. The summation in (5.68) is over all particles in the system. If there are many particles in the system, we can use an ensemble average $\langle ... \rangle$ when we carry out the summation, and thus (5.68) can also be written in the following form:

$$\sigma_{\alpha\beta} = -\frac{1}{V}\sum_{n<m}\langle f_{nm\alpha}R_{nm\beta}\rangle - P\delta_{\alpha\beta}. \tag{5.69}$$

If we use the bead–spring model for the polymer, then $f_{nm}$ only acts between neighbouring segments along the chain. This force is given by

$$f_{n,n+1} = \frac{3k_B T}{b^2}\frac{\partial R(n,t)}{\partial n}. \tag{5.70}$$

Thus, since $R_{n,n+1} = -\partial R_n / \partial n$, eqn (5.69) can be written as

$$\sigma_{\alpha\beta} = \frac{c}{N} \frac{3k_B T}{b^2} \sum_n \left\langle \frac{\partial R_\alpha(n,t)}{\partial n} \frac{\partial R_\beta(n,t)}{\partial n} \right\rangle - P\delta_{\alpha\beta}. \tag{5.71}$$

Here $c$ is the number density of the segments, and $N$ is the number of segments in one polymer. In (5.71), we have used the fact that there are $cV/N$ polymer molecules in the volume $V$.

In (5.71), we have only considered those forces that act between neighbouring segments along the chain. In concentrated polymer solutions, where the polymers are entangled and experience strong interactions, this may seem to be a very rough approximation. However, as we saw in Chapter 3, excluded volume or nematic interaction effects between polymers do not make a significant contribution to the stress. Further, although entanglement interactions have a large effect on the mechanical properties, this only changes the distribution function in (5.71) and there is no need to add an extra term to the equation for the stress. Therefore, (5.71) still holds true for concentrated polymer solutions.

Equation (5.71) shows that the stress of a polymer solution is determined by the directions of the segments. In actual fact, as can be seen from Fig. 3.3 or Fig. 5.8, the stress optical law for polymer solutions holds for a very wide range of the non-linear region. Even though the relationship between the velocity gradient and the stress can be very complicated, the linear relation between the stress and the birefringence holds precisely because of the fact that the stress is directly related to the molecular orientation.

### 5.3.3 The Rouse model

According to the picture outlined in Section 5.2, if the size of the polymer does not exceed the characteristic size $a$ of the tube, the polymer experiences no constraints from the tube, and so can be described in terms of the Rouse model. Actually, it is known that if the molecular weight of a polymer melt system is less than several hundred, then the diffusion or viscoelastic behaviour is well represented by the Rouse model (Fig. 5.4). Let us first investigate the viscoelastic properties of a polymer melt using the Rouse model.

Assume that the polymer is placed in the following flow field:

$$\bar{v}(r, t) = \kappa(t)r. \tag{5.72}$$

In this case, the segment $n$ has its velocity increased by an amount $\kappa \cdot R_n$, and so the Langevin equation (4.38) takes the following form:

$$\frac{\partial R_n}{\partial t} = \frac{k}{\zeta} \frac{\partial^2 R}{\partial n^2} + g(n, t) + \kappa \cdot R_n. \tag{5.73}$$

Rewriting (5.73) using the normalized coordinates introduced in Section 4.2.1 gives us

$$\frac{dX_p}{dt} = -\frac{X_p}{\tau_p} + \boldsymbol{\kappa} \cdot X_p + g_p. \tag{5.74}$$

Here $\tau_p$ is as defined in (4.46).

On the other hand, using normalized coordinates the stress is written as follows:

$$\sigma_{\alpha\beta} = \frac{c}{N} \sum_p k_p \langle X_{p\alpha} X_{p\beta} \rangle - P\delta_{\alpha\beta}. \tag{5.75}$$

Let us consider the case of shear flow. Here (5.74) becomes the following:

$$\frac{dX_{px}}{dt} = -\frac{X_{px}}{\tau_p} + \dot{\gamma} X_{py} + g_{px} \tag{5.76}$$

$$\frac{dX_{py}}{dt} = -\frac{X_{py}}{\tau_p} + g_{py} \tag{5.77}$$

$$\frac{dX_{pz}}{dt} = -\frac{X_{pz}}{\tau_p} + g_{pz}. \tag{5.78}$$

Further, the shear stress $S_{pxy}(t) \equiv \langle X_{px}(t) X_{py}(t) \rangle$ becomes

$$\sigma_{xy} = \frac{c}{N} \sum_p k_p S_{pxy}. \tag{5.79}$$

To calculate $S_{pxy}$, we multiply (5.76) and (5.77) by $X_{py}$ and $X_{px}$, respectively, and add the results. Taking an average gives us

$$\frac{d}{dt} S_{pxy} = -\frac{2S_{pxy}}{\tau_p} + \dot{\gamma} \langle X_{py}^2 \rangle. \tag{5.80}$$

If we consider weak flows, so that we can neglect terms with $\dot{\gamma}$ raised to an exponent of 2 or larger, we can replace $\langle X_{py}^2 \rangle$, the second term on the right hand side of (5.80), by its value when there is no flow, $k_B T / k_p$. Doing so, (5.80) becomes

$$\frac{d}{dt} S_{pxy} = -\frac{2}{\tau_p} S_{pxy} + \dot{\gamma} \frac{k_B T}{k_p}. \tag{5.81}$$

Solving this gives us

$$S_{pxy} = \int_{-\infty}^{t} dt' \frac{k_B T}{k_p} \exp[-2(t - t')/\tau_p] \dot{\gamma}(t'). \tag{5.82}$$

Substituting (5.82) into (5.75) gives us the following equation for the stress:

$$\sigma_{xy}(t) = \int_{-\infty}^{t} dt' G(t - t') \dot{\gamma}(t'). \tag{5.83}$$

Here, $G(t)$ is given by the following:

$$G(t) = \frac{c}{N}k_B T \sum_{p=1}^{\infty} \exp(-2t/\tau_p) = \frac{c}{N}k_B T \sum_{p=1}^{\infty} \exp(-2tp^2/\tau_1). \qquad (5.84)$$

Thus, the viscosity is calculated to be

$$\eta = \int_0^{\infty} dt\, G(t) = \frac{c\zeta Nb^2}{36}. \qquad (5.85)$$

This is proportional to the molecular weight. Actually, as shown in Fig. 5.10, for molecular weights that are not too large, the viscosity of polymer melts is indeed proportional to the molecular weight.

For the Rouse model, (5.74) can be solved to give the general relation between stress and strain. In this case, it can be shown that the steady shear viscosity is a constant independent of the shear rate.

### 5.3.4 The reptation model

Next, let us try to calculate the viscoelasticity for high molecular weight polymers using reptation theory. For simplicity, let us assume $Z \gg 1$ and ignore the length fluctuations of the polymer. Since the polymer segments are uniformly distributed along the tube, we can write $\partial R_n/\partial n = (L/N)u_n$.

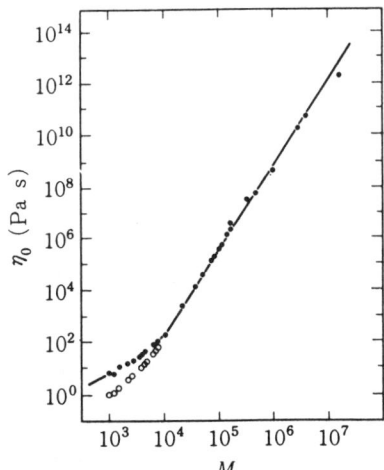

**Fig. 5.10** The relationship between viscosity and molecular weight for a polybutadiene melt. The black circles are the converted values of viscosity assuming identical free volume. The slopes of the straight sections are 1 and 3.4. (Colby, R.H., Fetters, L.J., and Graessley, W.W., (1987). *Macromolecules*, **20**, 2226, Fig. 5.)

Here $\boldsymbol{u}_n$ is the unit vector pointing in the direction of the tube at the position of the segment $n$. Thus (5.71) can be written as follows:

$$\sigma_{\alpha\beta} = \left(\frac{L}{N}\right)^2 \frac{3k_{\mathrm{B}}T}{b^2} \frac{c}{N} \sum_n S_{n\alpha\beta}(t) - P\delta_{\alpha\beta}. \tag{5.86}$$

Here

$$S_{n\alpha\beta}(t) = \langle u_{n\alpha} u_{n\beta} - \frac{1}{3}\delta_{\alpha\beta}\rangle \tag{5.87}$$

is the orientational order parameter tensor of the central axis of the tube.

To simplify matters, let us consider the following stress relaxation experiment. At time $t = 0$, we apply to a polymeric liquid at equilibrium an instantaneous strain, such that a point at $\boldsymbol{r}$ is moved to $\boldsymbol{r}' = \boldsymbol{E} \cdot \boldsymbol{r}$, and we maintain this constant strain while we measure the relaxation of the stress.

To calculate the stress, we must know how the central axis of the tube deforms under a macroscopic deformation. Here we will use the same assumptions as in Section 3.3.3:

(1) when the system undergoes a macroscopic deformation, the central axis of the tube undergoes an affine deformation;

(2) however, after a time $\tau_{\mathrm{R}}$ the tube length returns to its original equilibrium length $Za$.

Since here we are considering the case where $Z \gg 1$, the time for the tube length to return to its equilibrium length $\tau_{\mathrm{R}}$ is much shorter than $\tau_{\mathrm{d}}$. Therefore, in the following we will assume that $\tau_{\mathrm{R}}$ is negligibly small.

The equation for the stress (5.86) when the tube has returned to its equilibrium value $Za$ becomes the following:

$$\sigma_{\alpha\beta} = \frac{3ck_{\mathrm{B}}T}{N_{\mathrm{e}}} \frac{1}{N} \sum_n S_{n\alpha\beta}(t) - P\delta_{\alpha\beta}. \tag{5.88}$$

Here we have set $N_{\mathrm{e}} = a^2/b^2$.

Under assumption (1), a section which was pointing in the direction $\boldsymbol{u}$ before the deformation will point in the direction $\boldsymbol{u}' = \boldsymbol{E} \cdot \boldsymbol{u} / |\boldsymbol{E} \cdot \boldsymbol{u}|$ after the deformation. Thus, the initial value for $S_{n\alpha\beta}$ is

$$S_{n\alpha\beta}(t = +0) = \left\langle \frac{(\boldsymbol{E} \cdot \boldsymbol{u})_\alpha (\boldsymbol{E} \cdot \boldsymbol{u})_\beta}{|\boldsymbol{E} \cdot \boldsymbol{u}|^2} \right\rangle_0 - \frac{1}{3}\delta_{\alpha\beta}$$

$$\equiv Q_{\alpha\beta}(\boldsymbol{E}). \tag{5.89}$$

Here $\langle \cdots \rangle_0$ is as defined by (3.37).

Now, for times $t > 0$, the polymer gradually moves out of the stretched tube by reptation. For $t > 0$ the changes in $S_{n\alpha\beta}(t)$ are determined only by reptation, since there is no external flow. Thus, in a similar way to (5.41), the quantity $S_{n\alpha\beta}(t)$ is determined by the following:

$$\frac{\partial S_{n\alpha\beta}}{\partial t} = \frac{D_{\mathrm{t}}}{a^2} \frac{\partial^2 S_{n\alpha\beta}}{\partial n^2}. \tag{5.90}$$

At the boundaries $n = 0, N$, the newly created tube has an isotropic distribution. Thus we must have at $n = 0, N$ :

$$S_{n\alpha\beta} = 0. \tag{5.91}$$

Solving (5.89)–(5.91) gives us

$$S_{n\alpha\beta} = \sum_p^{\infty} \sin\left(\frac{np\pi}{N}\right) \exp\left(\frac{-tp^2}{\tau_d}\right) \frac{4Q_{\alpha\beta}(E)}{p\pi}, \tag{5.92}$$

where $p$ ranges over positive odd integers.
Thus the stress is

$$\sigma_{\alpha\beta}(t) = \frac{3ck_BT}{N_e} Q_{\alpha\beta}(E)\psi(t) - P\delta_{\alpha\beta}. \tag{5.93}$$

Here, $\psi(t)$ is as defined in (5.46).

In particular, under a shearing deformation with shear strain $\gamma$, we have for $\gamma \ll 1$,

$$Q_{xy}(\gamma) = \frac{\gamma}{5}. \tag{5.94}$$

Thus, from (5.93) we have

$$\sigma_{xy} = \frac{3ck_BT}{5N_e}\gamma\psi(t), \tag{5.95}$$

and so the stress relaxation function becomes

$$G(t) = \frac{3ck_BT}{5N_e}\psi(t). \tag{5.96}$$

Thus the viscosity is given by

$$\eta = \int_0^{\infty} dt G(t) = \frac{\pi^2}{20} \frac{ck_BT}{N_e}\tau_d. \tag{5.97}$$

Since $\tau_d \propto N^3$, this calculation shows the viscosity is proportional to the third power of the molecular weight.

On the other hand, according to experiments on polymer melts, the viscosity is proportional to the 3.4th power of the molecular weight $M$ (Fig. 5.10). As explained in Section 5.2.6, this discrepancy can be explained if we take into account the effects of the length fluctuations of the polymer along the tube.

If the strain $\gamma$ is not small, the shear stress can be written from (5.93) and (5.96) as follows:

$$\sigma_{xy}(t, \gamma) = \gamma G(t)h(\gamma). \tag{5.98}$$

Here

$$h(\gamma) = \frac{5}{\gamma}Q_{xy}(\gamma). \tag{5.99}$$

According to (5.98), the stress can be written as a product of a function of the strain and a function of time.

The above theory describes the actual features of stress relaxation quite well. Fig. 5.11 shows the non-linear stress relaxation function

$$G(t,\gamma) = \frac{\sigma_{xy}(t,\gamma)}{\gamma}. \tag{5.100}$$

for a polystyrene melt when a shear strain of $\gamma$ is applied, plotted against various strains. Under large strains, $G(t,\gamma)$ shows relaxation at two points: one at several tens of seconds and another at several hundred seconds. According to the theory presented above, these correspond to the process of a stretched tube returning to its equilibrium length, and the process of a polymer escaping from a deformed tube due to reptation. The respective relaxation times are $\tau_R$ and $\tau_d$. Actually, the ratio of the two relaxation times of $G(t,\gamma)$ is proportional to $Z^{1.5}$, which further supports this interpretation.

In Fig. 5.11, the curves of $G(t,\gamma)$ all have the same shape for $t > \tau_R$. In actual fact, these curves can be superimposed by shifting them along the vertical axis. This corresponds to the fact that the stress can be written as a product of a function of time and a function of strain for $t > \tau_R$ as in (5.98). We can thus calculate $h(\gamma)$ from the degree of shift along the vertical axis.

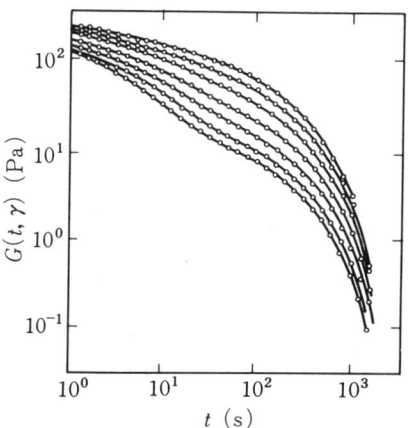

**Fig. 5.11** The non-linear stress relaxation function in a polystyrene solution. The molecular weight of the polystyrene is $8.42 \times 10^6$ and its concentration is $0.06\,\mathrm{g\,cm^{-3}}$. The uppermost curve shows the elastic relaxation modulus of linear viscoelasticity, and the remaining curves show stress relaxation in response to a finite shear strain. The size of the shear strain is, from above, 1.25, 2.06, 3.04, 4.0, 5.3, and 6.1. (Osaki, K., Nishizawa, K., and Kurata, M. (1982). *Macromolecules*, **15**, 1068, Fig. 1.)

Figure 5.12 shows $h(\gamma)$ determined from experiments compared with $h(\gamma)$ calculated from theory. Despite the fact that there are no parameters in the theoretical curve, there is good agreement with the experimental curve.

It is possible to expand the above calculation to general flow fields, leading to the following constitutive equation:

$$\sigma_{\alpha\beta} = 5 \int_{-\infty}^{t} dt'\, G(t - t')Q_{\alpha\beta}(E(t, t')). \tag{5.101}$$

Here $E(t, t')$ is the deformation gradient tensor at time $t$, based on the state at time $t'$. Equation (5.101) describes the non-linear viscoelastic behaviour of concentrated polymer solutions very well.

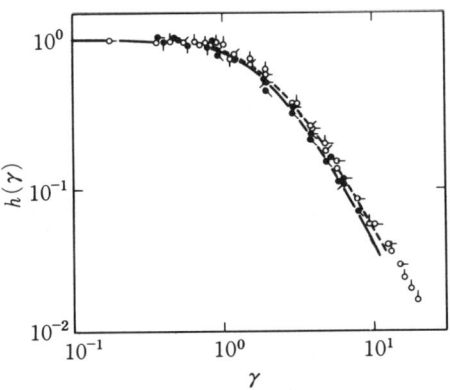

**Fig. 5.12** The behaviour of the function of strain $h(\gamma)$ component to the non-linear stress relaxation. The circles are experimental data, and the solid line is the theoretical curve. (Osaki, K., Nishizawa, K., and Kurata, M. (1982). *Macromolecules*, **15**, 1068, Fig. 3.)

# Bibliography

## General references on polymer physics

There has been a steady growth of the literature on polymer physics in recent years.
A classic work written by one of the founders of modern polymer physics is
  Flory P.J. (1953). *Principles of polymer chemistry*. Cornell University Press.
  This is a voluminous work which is now rather dated, and so may not be appropriate as an introductory textbook. However, if one reads it after becoming familiar with the basics of polymer physics, some fascinating insights can be found.
A work by a scientist who has revolutionised polymer physics is
  de Gennes, P.G. (1979). *Scaling concepts in polymer physics*. Cornell University Press.
  This elegantly-written book is widely read by both theoreticians and experimentalists, and has already become a classic in the field.
A book written with the emphasis on the dynamics of polymers is
  Doi, M. and Edwards, S.F. (1986). *The theory of polymer dynamics*. Oxford University Press.
  This work explains the calculations in some detail, and it is recommended to the reader of the current book who seeks more information.
A work on the new theories of polymer solutions, written by an expert on renormalization theory and an expert on neutron scattering experiments, is
  des Cloizeaux, J. and Jannink, G. (1990). *Polymers in solution, their modelling and structure*. Oxford University Press.
  This contains over 900 pages, and has a wealth of detail on both the renormalization calculations and the experiments.
Finally, an excellent work spanning the classical theories up to the modern techniques is
  Fujita, H. (1990). *Polymer solutions*. Elsevier, Amsterdam.

## Supplementary references

As well as the book by des Cloizeaux and Jannink, the following also explain the application of renormalization group theory to polymers:

Oono, Y. (1985). *Adv. Chem. Phys.*, **61**, 301.

Freed, K.F. (1987). *Renormalization group theory of macromolecules*. John Wiley.

An article detailing the statistical mechanics of polymers, with particular emphasis on the coil-globule transition, is

Lifshitz, I.M., Grosberg, A., and Khokhlov, A.R. (1978). *Rev. Mod. Phys.*, **50**, 683.

Brownian motion and the theory of stochastic processes are treated in

van Kampen, N.G. (1981). *Stochastic processes in physics and chemistry*. North-Holland, Amsterdam.

The classic theories of excluded volume effects and transport phenomena are fully treated in

Yamakawa, H. (1971). *Modern theory of polymer solutions*. Harper and Row, London.

A well-written textbook that covers the viscoelasticity and rheology of polymer systems is

Bird, R.B., Armstrong, R.C., and Hassager, O. (1987). *Dynamics of polymeric liquids*, Vols. 1 and 2. (2nd edn.) John Wiley, New York.

Finally, a book that details the constitutive equations of non-linear viscoelasticity is

Larson, R.G. (1988). *Constitutive equations for polymer melts and solutions*. Butterworth, Guildford.

# Index